UNDERSTANDING
THE HUMAN
BODY

UNDERSTANDING THE HUMAN BODY

Espen Dietrichs
Petter Hurlen
Kari C. Toverud
Peter Harrison (editor)

 CAMPION PRESS, EDINBURGH, 1992

British Library Cataloguing-in-Publication Data

Dietrichs, Espen
Understanding the Human Body
I. Title
612

ISBN 1 873732 05 8

English translation:	*Sally Hayward, MITI, MIL*
Cover and layout:	*Ellen Larsen*
Cover illustration:	*Kari C. Toverud*
Drawings:	*Kari C. Toverud*

Photos: p. 7,65,81	*Samfoto, Svein Erik Dahl*
p. 9,13,56	*Samfoto, John Petter Reinertsen*
p. 20	*Arnold Foerster, Rikshospitalet*
p. 24	*Gunnar Lothe, Anatomisk Institutt, Oslo* (tonsils)
p. 25,46	*Bjørn Hurlen, Odontologisk fakultet, Oslo* (white blood corpuscles, teeth)
p. 29	*Sven Mjøern, ØNH-avd., Ullevål sykehus* (vocal cord)
p. 32,33	*Gunnar Lothe*
p. 46,47	*Steinar Aase, Rikshospitalet* (tongue, stomach)
p. 50	*Smith, Kline & French, Norway*
p. 58,59	*Scandecor International AB* (skeletons)
p. 59	*Røntgen, Oslo kommunale legevakt*
p. 63	*VG-foto, Tore Berntsen* (footballers)
p. 75	*Lennart Nilsson*
p. 77	*Samfoto, Rune Lislerud*
Other photos:	*Stein Grebstad*
Lithography:	*Scan Lith, Copenhagen*
Printing:	*Emil Moestue a.s.*

First published in 1992   ISBN 1 873732 05 8

© 1988 UNIVERSITETSFORLAGET AS I OSLO

Den Forunderlige Kroppen (The Amazing Body)

© 1992 CAMPION PRESS LIMITED
384 Lanark Road
Edinburgh EH13 0LX

Contents

Introduction

The study of anatomy and physiology is not a new academic discipline. It is an area of study however that is constantly evolving and updating due to new discoveries within medicine and its allied fields. These areas of study were primarily used in the training of medical, paramedical and nursing personnel although more often this subject material is now part of the core knowledge for students training for careers in life science, health studies, sport studies and physical education. These courses are now being taught outside the traditional medical and nursing schools in colleges of higher and further education and tertiary colleges.

This book has been written to assist students working in schools and colleges who are taking A level courses in life sciences or who are embarking on nursing or health studies courses. It is suitable as a course text book or as a reference book in the department, school or college library.

The book adopts a unique approach in that it attempts to integrate text, human photographs and illustrations in such a way that it makes the material easier for the reader to absorb, understand and subsequently remember. The phrase 'a picture tells more than a 1000 words' is most appropriate as the underlying philosophy of this book.

It is not the intention for the pictures and illustrations to replace text material; in this book they are used rather to reinforce facts and give greater depth thus stimulating further reading. Where possible actual photographs of the body have been used along with superimposed illustrations to give greater realism. Complex topics are dealt with in a simple comprehensible manner and key points and facts are highlighted in each section making the text suitable for revision purposes. In this way the reader will be able to absorb more practically-based information in an easy-to-understand manner in a relatively short period of time.

The book is not divided into traditional headings. Each subject is dealt with over two pages. Under each subject, the text in the blue panel (like this) refers to other related subjects which you will also find useful to study.
The figures in brackets (00) refer to the relevant pages.

In the red box a short description is given of some of the most common or important diseases, or injuries which are related to the material dealt with on that page.

About the book

The book has a *lexi-visual format* with the text and illustrations closely linked together. This style is especially appropriate for the study of anatomy and physiology and the format requires a particularly high standard of design. The book contains 40 separate subject areas and the cohesion between the text and the illustrations is extremely important for the understanding of the subject matter. To achieve this aim the authors, illustrators and designers have worked together as a team and have each made a significant contribution to the final format of the book. This has resulted in a balanced and interesting style, in which the outlines of the structure of the human body are used as a foundation for the understanding of the most important physiological mechanisms and processes.

In order to keep the subject matter as uncomplicated as possible, only the most important names and terms from anatomy and physiology have been used. At the back of the book are some tables and more detailed anatomical terms for those who are interested

We hope that you enjoy your journey of discovery through **Understanding the Human Body.**

Summer 1992

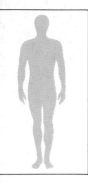

Anatomy and physiology

Anatomy is the science of the structure of the human body. *Physiology* is the science of the different functions of the human body.

The human body

The human body is a fantastic structure. In adults it is composed of at least 100,000,000,000 (one hundred billion) tiny living *cells*, which all originated from a single fertilized cell. The first cells all look alike to start with, then they begin to develop in different ways, so that different types of cells come into being. Thus they form the skeleton, the muscles, the intestines, the skin, the hair and everything else of which a body is made up. Although the different types of cells have specific tasks, they all work together to make the body function efficiently.

Characteristics of life

The different tasks of the cells are divided into:
- taking in nourishment (absorption);
- breathing (respiration);
- combustion of nutrients (metabolism);
- growth and reproduction;
- production of useful materials (secretion);
- elimination of waste (excretion);
- sensitivity to stimuli;
- transmission of impulses;
- contractions (to give movement).

In the case of single-cell *organisms* , one individual cell has all these characteristics. With multi-cell animals however the tasks are divided up among the various types of cells and this is how it is with humans. Cells possess one or several characteristics but are lacking others. Nerve cells, for example, cannot reproduce themselves. But the intake of nourishment, respiration and combustion are such important features that all cells are dependent on them in order to live.

The cell

Our bodies, like all living organisms, are made up of living cells, building materials and nutrients. There are many different types of cell present in the body. The largest can have a diameter of 0.2 millimetre, the smallest merely 0.002 millimetre. Yet all cells have common features.

See also:
Chromosomes and cell division (12, 13)

For more about nutrients, see:
The digestive system (44, 45)

The cell structure

The cell is surrounded by a film called the *cell membrane*. This consists principally of fat. Inside the cell membrane are the *cytoplasm* and the *cell nucleus*.

The cytoplasm is mainly *water* plus *salts* (e.g. common salt), *minerals* (e.g. iron) and *nutrients* and *building materials* (*fats, carbohydrates, proteins* (e.g. albumen) and *nucleic acids*). Also in the cytoplasm, there are a number of minute forms with specific tasks. These we call *organelles*.

Transport in and out of the cell

The cell membrane has *pores* and controls what materials pass in and out through these pores. Some materials, such as water and chloride, automatically flow from places where there is too much to places where there is too little. This process is called *diffusion*.

Other materials, such as amino acids and sodium, have to be conveyed through the pores by a positive pumping mechanism. This kind of transport requires *energy*.

Furthermore, the cell can take in or eliminate larger particles, mainly waste, by wrapping itself round these particles and sealing off a piece of the membrane.

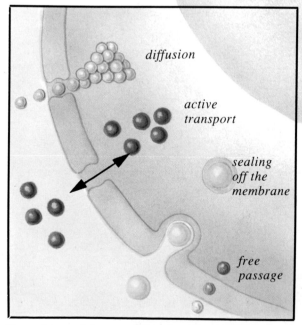

diffusion

active transport

sealing off the membrane

free passage

Fat and materials containing fat can pass unhindered through the membrane.

he cell nucleus

he *cell nucleus* also has a film covering, the
uclear membrane. Inside the nucleus are long
rings of *nucleic acids* (DNA, *deoxyribo-nucleic
cids*). These we call *chromosomes*.

Composition of proteins

The chromosomes govern the production of proteins
and determine how these are made up. All proteins
consist of amino acids. The number and the order of the
amino acids determine the properties of the protein.

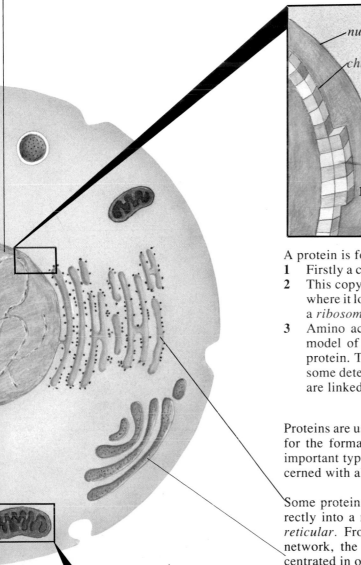

A protein is formed as follows:
1 Firstly a copy is made of a part of the chromosome.
2 This copy goes from the nucleus to the cytoplasm,
   where it lodges against a spherical *organelle*, called
   a *ribosome*.
3 Amino acids in the cytoplasm gather around the
   model of the copy chromosome and thus form a
   protein. The order of nucleic acids in the chromosome determines in what sequence the amino acids
   are linked together.

Proteins are used among other things as building bricks
for the formation and maintenance of cells. Another
important type of protein is the *enzyme*, which is concerned with almost all chemical reactions in the body.

Some proteins are secreted by the cell. They pass directly into a network of membranes, the *cytoplasmic
reticular*. From there they are conveyed to another
network, the *Golgi apparatus*, where they are concentrated in order to be sent out of the cell.

mitochondrion

Combustion

The cell requires *energy*, among other things, in
order to build materials and transport them and for
producing energy and heat. The cell obtains energy
through the combustion of nutrients. Combustion
mainly takes place in the *mitochondrions* and is
dependent on *oxygen* (O_2). By- products of this are
water (H_2O) and *carbon dioxide* (CO_2); these are
subsequently excreted by the cell.

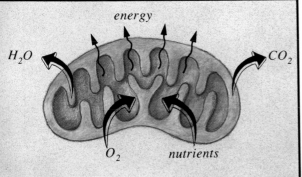

Chromosomes and cell division

The cell nucleus contains chromosomes which govern protein production. Many characteristics of the body are determined by the type of proteins which the cells make.

The cells in the body must be constantly renewed. New cells are created by a cell splitting into two new, identical cells.

Chromosomes

Chromosomes govern the production of proteins. The part of a chromosome that governs the production of a specific protein is called a *gene*. Each chromosome has many genes and looks after the production of many proteins.

Every cell in our body has 23 pairs of chromosomes (except for sex cells, see p. 73). The pairs of chromosomes are of the same type and have genes which control the production of the same proteins.

Two chromosomes form an exception to this rule. They are called X and Y. All cells in the female body have two *X-chromosomes*, whilst the cells in the male body have one *X-chromosome* and one *Y-chromosome*. Therefore the X and Y chromosomes are also called the *sex chromosomes*.

Gene→

The genes determine what th proteins are to be and thereby als the characteristics of the body.

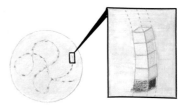

chromosome with a gene

See also:
The cell (10, 11)
Heredity and environment (78, 79)
Reproduction (73)

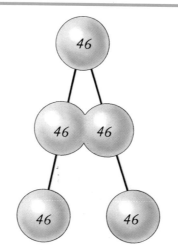

When a cell divides it is important that each new cell has a copy of all the genes. The two new cells must then each have 23 pairs of chromosomes. Therefore an identical copy of each chromosome is made before the cell divides.

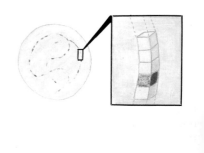

Another example of the conne tion between genes and chara teristics is sex. When the bo has two X-chromosomes, it d velops female sexual feature With one X and one Y chrom some, the body acquires ma sexual features.

Cell division (mitosis)

1 The cell before division. In order to keep the drawing simple, only 4 of the 46 chromosomes are illustrated.

2 Cell division starts by making an accurate copy of each chromosome, which produces double chromosomes.

ein → Feature

When the cells in the body have genes for making a specific protein which forms a lot of pigment in the iris, then people have brown eyes.

If the body cells have genes for making another protein which does not form much pigment, then the eyes are blue.

3 The nuclear membrane dissolves and disappears. The two identical chromosomes split apart and each moves to one side of the cell.

4 The cell divides into two. The cytoplasm spreads itself over the two new cells. Each cell has the same number of chromosomes as the original cell.

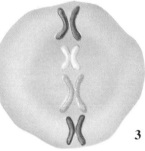

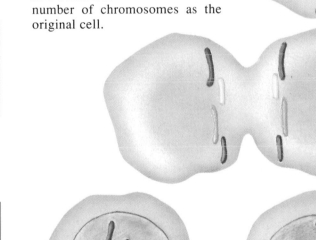

XY   *XX*

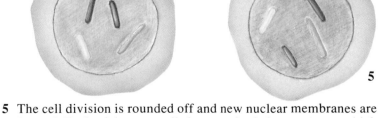

5 The cell division is rounded off and new nuclear membranes are formed. Each of the new cells has acquired chromosomes which are precisely identical to the chromosomes in the original cell.

13

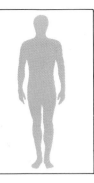

Tissue

All cells are basically identical and have the same properties but in the human body the cells have specific tasks. This influences both their appearance and their functioning. A large number of cells of the same type together forms tissue. The main groups are skin and glandular tissue, nerve tissue, muscular tissue and connective and supporting tissue.

Connective and supporting tissues

The connective and supporting tissues strengthen, protect and give shape to the body. They consist of a number of types of cell. Just as important as the cells is the *interstitial matter*, consisting of matter excreted by the cells and which lodges betwee them. It is the interstitial matter which gives th connective and supporting tissues their characte istic features.

Connective tissue is found in most parts of the body. The *connective tissue cells* make fibres, which form part of the interstitial matter and which make the connective tissue strong and elastic. This makes the connective tissue resistant to wear and tear and loads, and able to protect the other tissues. Some connective tissues are extra firm and strong, such as *tendons* and *ligaments*. Others are somewhat looser, such as in the deeper layers of the skin. A special type of connective tissue is the *adipose tissue*, which consists of *fat cells*.

Cartilage tissue is harder and firmer than connective tissue, but softer than bone tissue. It consists of *cartilage cells* and interstitial matter. There are different types of cartilage. One particularly smooth type is found on the *joint surfaces* where the bones are constantly sliding over each other. An elastic type of cartilage is in the ear and the nostril.

Bone tissue is to be found in all th bones of the skeleton. It is formed b the *bone cells* making fibres as pa of the interstitial matter. *Calcium* deposited between the cells and th fibres. This makes the bone tissu hard. The bone cells are constantl active and secrete calcium continu ously.

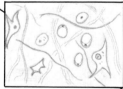

For more about
bones and cartilage,
see:

Lymph

Between the cells and the interstitial matter there is fluid. This is called lymph. There is a continuous exchange of matter between 1. the blood and 2. the lymph and the cells.

Other tissues

There are tissues which are difficult to classify under one of the four main groups. These are, for example, *blood and blood-producing tissue*, *sensory tissue* and *sex cells*.

Skin and glandular tissue

Skin and glandular tissue consists of *epithelial cells*. The epithelial tissues have two main tasks. Some epithelial tissues cover surfaces, whilst others control *secretion* (production of materials necessary for the body).

Epithelial tissue has a variety of appearances and properties, depending on the surface it covers.

We find *different layers of epithelial cells* on surfaces with a lot of wear and tear and on surfaces which matter must not penetrate. The skin consists of such epithelium. It is found moreover in the mouth cavity, the throat and in the vagina.

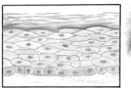

One layer of *flat epithelial cells* gives a smooth surface and lets matter through easily. We find it mainly in three places: on all surfaces of the body's orifices, where little friction occurs; on the inside of the blood vessels, where water and dissolved matter can easily pass in and out of the blood; in the lungs where gases are exchanged between the air and blood.

A layer of *thick epithelial cells* is found mainly where the body controls the passage of matter, for example on the surfaces of the *mucous membrane* on the inside of the stomach.

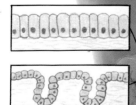

Epithelial tissue which specializes in secretion is called *glandular tissue*. Such tissue is found in the *digestive glands* (e.g. *salivary, gastric and intestinal glands*) and in the *hormone-producing glands*.

Nerve tissue

Nerve tissue consists principally of *nerve cells* or *neurones*. The neurones have the task of conducting *electrical signals (impulses)*. They have offshoots called *axons* and *dendrites* Fully developed nerve cells are unable to divide. After reaching the adult state, division comes to a halt. Shortly after birth, no new nerve cells are formed.

The short nerve fibres which conduct the impulses to the cell body are called *dendrites*, whilst the *axons* carry the impulses away from the cell body. The axon is surrounded by a *medullary sheath*, a layer made of connecting tissue cells. The medullary sheath ensures that the impulses are conducted more rapidly.

Muscular tissue

Muscular tissue consists of *muscle cells*. The muscle cells can contract. There are three types of muscular tissue.

Striated muscular tissue is found in the skeletal or voluntary muscles. The name derives from the characteristic stripes. The muscle cells are normally 1 to 50 mm long but may be as long as 40 cm. They are referred to as muscle fibres. The contractions in the striated muscular tissue are under voluntary control.

The *cardiac muscular tissue* is only found in the heart. It has the same kind of stripes as the striated muscular tissue but the similarity stops there. The cardiac muscle cells are short and do not form fibres. The cardiac muscle tissue is not under voluntary control.

Smooth muscular tissue is mostly found in intestines. The muscle cells are elongated and not striated. Contraction of the smooth muscular tissue is automatically controlled by the autonomic nervous system and is not under voluntary control.

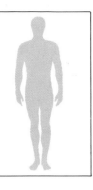

The organ systems

The organs are made from different types of tissue. The brain, the stomach and the kidneys are examples of organs. Within the body the different organs work together on the various tasks. We can distinguish eleven organ systems which work together. Each of these organ systems has different functions.

See also:
Coordination of the body systems at rest (82, 83)
Coordination of the body systems during physical exertion (84, 85)

The skin

consists of the *epidermis, dermis, subcutaneous fat*, hair, nails, sebaceous glands, sweat glands and mammary glands. The skin protects the other organ systems against both damage from outside and from dehydration.

The skeleton

consists of bones and the connections between them. A moveable connection is called an *articulated joint*. The skeleton keeps the body upright and protects the internal organs.

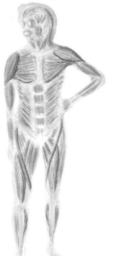

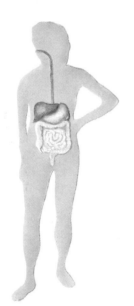

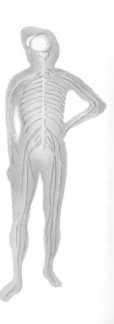

The nervous system

consists of the brain, the spinal cord, the nerves, the nerve ends and the sense organs. It also directs and co-ordinates the other organ systems.

The digestive system

consists of the digestive tract (from the mouth to the anus) and glands which produce saliva and digestive juices. These ensure that food is broken down, so that the nutrients are absorbed through the stomach wall.

The muscular system

consists of muscles and tendons which attach the skeletal muscles to the bones. We are able to move because the muscles are able to contract.

The lymphatic system

consists among other things of the spleen, the thymus, the tonsils, the lymph nodes and the lymph vessels. It is important for the defence of the body against diseases.

The respiratory system

consists of the air passages (nose, throat, larynx, windpipe (trachea) and bronchial tubes) and the lungs. When we breathe, oxygen is conveyed from the air into the blood and carbon dioxide from the blood to the air.

The urinary system

consists of the kidneys, the ureters, the bladder and the urethra. This looks after the removal of water and dissolved waste matter.

The endocrine system

consists of glands which secrete hormones. Hormones are stimuli which are conveyed round the whole body by the blood.

The cardio-vascular system

consists of the heart and blood vessels. It carries gases, nutrients and waste through the body.

The reproductive system

consists of the sex organs. It controls reproduction.

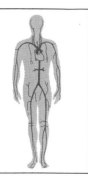

Heart and blood vessels

The heart, blood vessels and blood form the circulation or transport system within the body This system ensures that oxygen, nutrients and water are conveyed to all the tissues in the body. It also ensures the removal of carbon dioxide and waste matter from the tissues. In addition, white corpuscles, hormones and other materials are transported.

The blood vessels

The blood flows through small pipes, the *blood vessels*. The heart pumps the blood through these vessels. The blood vessels consist of connective tissue and smooth muscular tissue. The inside is covered with a thin layer of smooth epithelium.

From the heart the blood is pumped through a network of blood vessels, the *arteries*. The blood first reaches the main arteries. They divide into smaller branches (*arterioles*). In this way the blood is finally spread throughout the whole body via the arteries.

From these arteries the blood passes into an extensive network of *capillaries* which are found in the tissues. Altogether the body has about 100,000 kilometres of capillaries. The capillaries are thin, so that fluid, matter and gases can easily be exchanged between the blood and the surrounding tissue.

The blood is once more drawn into small blood vessels called *venules*. These converge into larger *veins* and finally into a few main veins, whence the blood flows back to the heart.

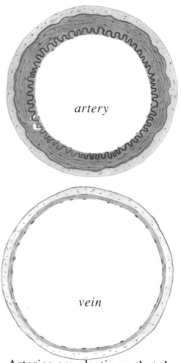

artery

vein

Arteries are elastic so that they can dilate each time the heart pumps out blood. The veins resemble the arteries, but have a thinner wall. Blood pressure in the veins is lower. The veins may have *valves*, which ensure that the blood can only flow one way. When we use our leg muscles, they exert pressure on the veins and drive the blood upwards.

When we stand still for a long time, we run the risk of too much blood accumulating in the veins of the legs. This can cause the blood supply in the rest of the body to become inadequate and we start to feel unwell. It is sensible to use the leg muscles a little when we have to remain standing in the same place for any length of time.

For more information about gas exchange, see: The respiratory system (26, 27)

See also: The blood (22, 23) The heart (20, 21)

For the names of the most important blood vessels, see: Anatomical terms (88)

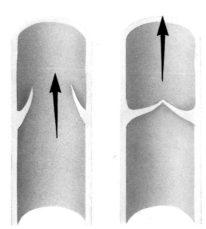

the working of the valves in the veins

The greater and lesser circulation

The greater and lesser circulatory systems are two different parts of the the blood circulation. They are connected to each other by the heart. The blood vessels from the heart to the lungs, the blood vessels in the lungs and the blood vessels from the lungs back to the heart belong to the *lesser* or *pulmonary circulation*. In the lungs the blood absorbs oxygen (O_2) and gives off carbon dioxide (CO_2).

pulmonary artery

pulmonary vein

the lesser circulation

All other blood vessels belong to the *greater circulation*. In the tissue the blood gives off O_2 and absorbs CO_2.

vein

artery

capillaries

All the blood from the greater circulation reaching the heart is pumped to the lungs via the lesser (pulmonary) circulation. The blood reaching the heart from the lesser circulation is pumped into the greater (systemic) circulation. In this way, the blood circulates in turn through the lesser and the greater circulation. Oxygen-rich blood is bright red, oxygen-starved blood is dark red.

Diseases

Varicose veins occur particularly in the veins which lie close to the skin of the legs. If a valve in a vein collapses, the pressure on neighbouring valves becomes greater. That can cause the vein to dilate and a varicose vein occurs.

Nicotine in tobacco smoke can make the small blood vessels in the arms and legs contract. Then the blood supply deteriorates. With long-term smoking this can lead to the destruction of tissue.

With *ageing*, fatty matter can be deposited on the walls of the arteries. This fat can harden and we call this *hardening of the arteries*. Hardening of the arteries makes the blood vessels narrower and less elastic. This can lead to the blood supply to the body's organs being reduced. It is thought that certain eating habits can result in hardening of the arteries.

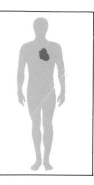

The heart

The heart has the task of pumping the blood round the body. The heart is the size of a clenched fist and is located slightly left of centre in the thoracic cavity.

The structure of the heart

The heart is constructed from cardiac muscle tissue (*myocardium*). The outside of the heart is covered with a smooth membrane, the *epicardium*. The surrounding tissue is covered with the same sort of membrane. These two membranes form the *pericardium*. Between them there is a narrow, fluid-filled cavity. The pericardium and the fluid ensure that the heart slides smoothly against the tissue when it beats.

The cardiac muscle tissue obtains oxygen and nutrients through its own arteries. These are called *coronary arteries*.

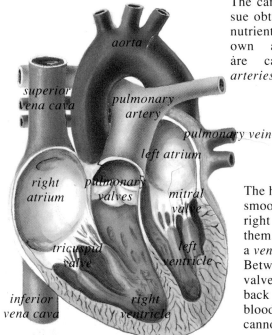

aorta

superior vena cava

pulmonary artery

pulmonary vein

left atrium

right atrium

pulmonary valves

mitral valve

tricuspid valve

left ventricle

inferior vena cava

right ventricle

The hollow cavities within the heart are covered with a thin, smooth membrane (*endocardium*). The heart is divided into a right section and a left section, with a dividing wall between them. Each half consists of two hollow cavities, an *atrium* and a *ventricle*.
Between each atrium and each ventricle there is a complex valve. This valve serves to prevent the blood from flowing back into the atrium. There are also valves located where the blood vessels leave the ventricle (aortic valves). The blood cannot then flow out of the arteries back to the ventricles.

Blood pressure

The pressure in the blood vessels is higher than in the surrounding tissue. This pressure is also higher in the arteries than in the veins, so that the blood always flows in one direction. We call the pressure in the arteries the blood pressure. This is higher when the ventricles contract (the *systole*) than when these are at rest (the *diastole*). Therefore we measure two blood pressures, the upper pressure (systolic pressure) and the lower pressure (diastolic pressure).
Blood pressure is measured with an inflatable sleeve which is wrapped around an arm or a leg. When inflated, the sleeve compresses the artery. As it is slowly released, the pressure can be read off from a mercury manometer. The upper pressure is read off as soon as blood flows through the artery again. The lower pressure is registered when there is no further interference with the blood flow from the (partially) inflated sleeve. Normal blood pressure is 120/80 mm Hg (mercury).

See also:
Heart and blood vessels (18, 19)
The blood (22,23)

When the heart functions

When the cardiac muscle tissue contracts, the heart operates like a pump and forces the blood forward. The valves determine which way the blood flows.

The atria and the ventricles take it in turns to contract and relax. One contraction and one relaxation are together called a pulse beat. The pulse beat is divided into two: the systole is the time during which the ventricles contract and remain contracted; the diastole is the time during which the ventricles relax and remain relaxed.

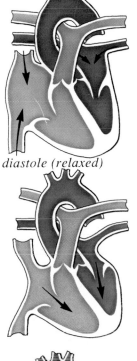

diastole (relaxed)

The blood first passes into the atria.

When the muscular tissue in the atria contracts, the blood is pumped into the ventricles The blood passes from the right atrium to the right ventricle and from the left atrium to the left ventricle.

Then when the muscular tissue in the ventricles contracts, the blood is pumped out of the right ventricle of the heart through the pulmonary artery to the lungs. From the left ventricle the blood is pumped through the aorta to the rest of the body.

systole (contracted)

The heart beats automatically and we are scarcely able to influence our heartbeats at will. At rest the pulse rate is 60-70 beats per minute (*heart frequency*). The heart pumps out 4-5 litres of blood per minute (*heart/minute-volume*); that is sufficient to cover the body's oxygen requirement. When we start walking, working or using our muscles in any other way, the body needs more oxygen. Then we start to breathe more quickly and the heart frequency increases in order to pump the blood round the body more rapidly. This is controlled by the cardio-respiratory centre in the brain. When we work hard, the pulse rate can rise to 200 beats per minute and the heart/minute-volume increases to over 20 litres.

Contraction of the atria and the ventricles is controlled by electrical impulses which resemble the stimuli in the nervous system. Impulses reach the heart at the sino-atrial node and then spread across the rest of the heart. This ensures that the cardiac muscles contract. The electrical impulses in the heart can be recorded on a special measuring instrument. The recorded trace is called an *electrocardiogram* (ECG).

The pulse wave

The connective tissue in the arteries is elastic. Each time the heart pumps out blood, the arteries dilate slightly and then contract again. When we feel an artery, we can detect this as a pulse wave.

Diseases

The coronary arteries may become narrowed, for example with hardening of the arteries. They can become so narrow that the heart does not get enough oxygen when it is working hard. That is terribly painful. The disease is called *heart spasm* or *angina pectoris*.

If one of the coronary arteries becomes completely blocked, the cardiac muscle tissue dies in that part of the heart that supplies the artery with blood. This is called a *cardiac infarct*. The dead muscle tissue is gradually replaced by connective tissue.

Occasionally blood pressure may become unnaturally high. When this happens, it is accompanied by unpleasant feelings and various organs in the body may later suffer damage. Frequently, there is no cause indicated for this complaint.

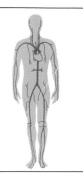

The blood

The blood has the function of transporting gases, matter and cells around the body. An adult human has 4-5 litres of blood, and slightly more than half the blood is fluid. This fluid is called plasma. The remainder consists of white and red blood corpuscles and blood platelets.

The red blood corpuscles

The main function of the red *blood corpuscles* (*erythrocytes*) is to transport oxygen from the lungs to the tissue in the body. There are about 5 million red blood corpuscles in one mm^3 of blood. The most important matter in the red blood corpuscles is *haemoglobin*. Haemoglobin is a large molecule, which has the ability to combine with oxygen. *Iron* is an important constituent of haemoglobin. The red blood corpuscles are some of the smallest cells in the body. They are formed in the *red bone marrow*. When they are almost fully formed, they lose their cell nucleus. Mature red blood corpuscles thus have no nucleus, unlike most other cells.

granulocyte

granulocyte

monocyte

flat epithelial cells in capillary wall

Plasma

Plasma is a fluid containing proteins. More than 90% of plasma is water. A number of materials are soluble in water and can therefore be easily transported through the body with the blood.

Electrolytes are examples of such materials. Sodium chloride (common salt) is the material most present as electrolyte. Sodium chloride is divided into free, charged atoms (ions): sodium (Na^+) and chloride (Cl^-). Potassium (K^+) and bicarbonate (HCO_3^-) are other important electrolytes. Bicarbonate has the ability to absorb or release free hydrogen ions (H^+), so that the blood can maintain the correct degree of acidity which is measured by the pH level. The normal pH level of blood is pH 7·4. Bicarbonate acts therefore as a *buffer*.

The remainder of the plasma consists principally of proteins – *plasma-proteins*. Most of them have the task of transporting different materials, such as hormones, fats, vitamins and iron.

Antibodies are a type of plasma-proteins which operate in the field of immunization. Other proteins play an important role in coagulating the blood in the event of injury (fibrinogen among others). In addition, certain proteins play an important role in filling the blood vessels. They attract fluid which would otherwise accumulate in the tissue.

For coagulation of the blood, see: The functions of the skin (56, 57)
For more about blood types, see: Heredity and environment (78, 79)
For more about white blood corpuscles and macrophages, see:
The lymphatic system (24, 25)
See also:
Heart and blood vessels (20, 21)

Blood platelets

The *blood platelets* (*thrombocytes*) are not true cells but small plates formed by the splitting of the large cells in the red bone marrow. Blood platelets are important for *coagulating* the blood in the event of injury. There are about 300,000 blood platelets in one mm³ of blood.

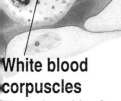

lymphocyte

granulocyte

White blood corpuscles

The *white blood corpuscles* (*leukocytes*) protect the body from infection. Each mm³ of blood contains about 7,000 white blood corpuscles. The main groups of white corpuscles are: *lymphocytes,* formed in the lymph nodes, *monocytes* and *granulocytes* produced in the red bone marrow. The monocytes can be converted outside the blood into *macrophages*. The white corpuscles contribute in a variety of ways towards combating infection. Some track down foreign organisms and cells and mark them so that others can kill or consume them.

Blood groups

There are large molecules on the surface of all cells. These are called *antigens*. The antigens on the surfaces of the red corpuscles can be divided into groups. Each group is associated with a specific blood group system. *AB0* (*AB-zero*) is one such group. Another is *Rhesus* (*Rh*). People with identical antigens within each system, have the same *blood group.*

The body can produce antibodies capable of combining with a specific antigen. For most blood group systems (Rhesus, for example) the body is able to make antibodies against a blood group other than its own. This can happen when the body comes into contact with blood of another type, such as with a blood transfusion and sometimes at birth. In the AB0 system, the body possesses all the antibodies for binding the antigens which the blood corpuscles in its own body do not have.

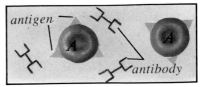

Blood group A is associated with one special antigen. Therefore, the plasma contains antibodies against blood group B.

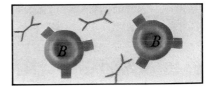

Blood group B is associated with another antigen and has antibodies against blood group A.

People with *blood group AB* lack antibodies against both blood group A and B.

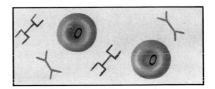

People with *blood group 0* have antibodies against both blood group A and B.

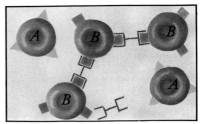

If a person of blood group A is given blood cells of type B, then the antibodies in the blood will ensure that the type B blood cells are made to coagulate. This can be very dangerous and often proves fatal.

Diseases

Anaemia (*haemoglobin deficiency*) is the collective name for situations where the blood has too little haemoglobin. The most common cause is a lack of iron. When the body has an iron deficiency, it cannot make sufficient haemoglobin. The haemoglobin content of the blood is measured by means of a blood test. The *sedimentation rate* is a blood test in which blood which has been rendered non-coagulating is measured to see how far the blood cells sink in a 20 cm long tube during the course of an hour. Normally they only sink a few millimetres. In many diseases they sink more rapidly.

The lymphatic system

The main task of the lymphatic system is to combat infection. It assists in the removal of dead cells and the transportation of fluid and fat, and consists of lymph, lymph vessels and lymph tissue.

The lymph

In the arterial capillaries the fluid is driven into the tissue by blood pressure. Most of the fluid is drawn back into the venous capillaries by osmotic pressure, but a small proportion still remains in the tissue. This fluid, which is conveyed by the lymph vessels, is called *lymph*. Lymph contains a great number of white blood corpuscles. Lymph from the intestines may contain a great deal of fat after a meal rich in fat.

The lymph vessels

Lymph is collected by a network of lymph vessels. The smallest resemble capillaries. They connect up with increasingly larger vessels, the *thoracic duct* among others. The largest lymph vessels syphon the lymph off into the main veins in the upper part of the thoracic cavity.

For more information
about white blood
corpuscles, see:
The blood (23)

Lymphatic tissue

Lymphatic tissue is found everywhere in the body. It is collections of lymphocytes, a type of white blood corpuscle, which lie in a network of connective tissue.

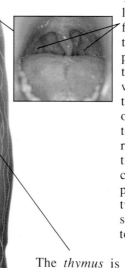

The tonsils are lymphatic tissue found in the throat. They try to prevent any infection from bacteria which enters through the mouth or the nose. The tonsils form a ring of lymphatic tissue. The ring consists of the pharyngeal tonsil, two palatine tonsils and the lingual tonsil.

The *thymus* is lymphatic tissue which lies in front of the heart. The thymus is large in children and becomes smaller after puberty and probably plays a large role in the development of the body's *defence mechanism.*

The *spleen* is the largest organ with lymphatic tissue. This organ participates in the defence mechanism by purging the blood of foreign micro-organisms and destroying them. The spleen also captures old 'burnt-out' red corpuscles and decomposes them.

The *lymph nodes* are encapsulated, small, oval structures of lymphatic tissue with lymphocytes and other white blood corpuscles. We find them mainly at points where a number of lymph vessels join. They filter the lymph and try to hold up foreign micro-organisms or matter, breaking them up and consuming them so that they cannot reach the blood.

The defence mechanism

The body's defence apparatus against infection and other foreign organisms and elements is called the *defence mechanism*. *Viruses*, *bacteria* or *waste* can penetrate through the body's natural orifices or through the mucous membranes and skin. The skin is, in normal circumstances, impenetrable for foreign organisms. When the skin is damaged, for example by a wound, bacteria are able to penetrate. The bacteria feed on body materials, multiply and try to penetrate further into the body. In order to be able to combat foreign organisms, the defence mechanism has to notice that they are there. The defence mechanism is able to register large molecules on surfaces, such as for example on the cell membrane of a bacterium. We call such molecules *antigens*. The defence mechanism is capable of distinguishing between antigens on its own cells and antigens on a foreign organism.

Tissue in which the defence mechanism is active, usually swells up and is painful. When this happens directly under the skin, the skin becomes red. We call this *inflammation*.

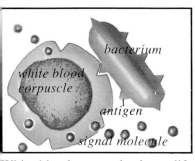

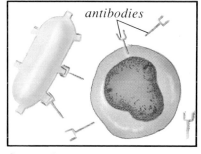

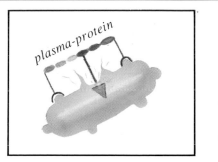

White blood corpuscles have different tasks within the defence mechanism. Some are able to track down foreign organisms. They are warned by minute *signal molecules*. When the bacteria damage the tissue, matter is released which acts as a signal molecule. Some ensure that white corpuscles are directed to the wound, whilst others send signals to the temperature centre in the brain so that the body temperature rises (fever).

Some lymphocytes, *B-lymphocytes*, mutate and start producing a certain type of protein which can attach itself to the antigens on the bacteria. We call such proteins *antibodies*. The antibodies indicate that something strange is happening. Some antibodies can render the bacteria harmless by making them coagulate.

Bacteria which have been 'marked' by antibodies can be destroyed by being punctured by a special type of plasma-protein.

Every type of white blood corpuscle has the task of killing foreign organisms and some even have to 'eat' them (macrophages among others). Individual T-lymphocytes can kill foreign organisms which have not been marked beforehand.

Frequently an infection is stopped at the point where the bacteria penetrated. Sometimes the infection spreads further into the body. Some bacteria penetrate the small lymph vessels and are conveyed directly to the lymph nodes with many white blood corpuscles. This activates the lymphocytes which start dividing, so that there are more of them to withstand the attack. Therefore, the lymph nodes swell up and become painful. The same thing happens with the tonsils in a throat infection. When the infection has been overcome, certain lymphocytes become special *memory cells*. These are able to set the defence mechanism into operation particularly quickly when the body is again infected by the same micro-organisms. As a rule, the infection is then stopped before we become ill. That is why, for example, we usually get childhood illnesses only once. By *vaccination*, dead or very weak micro-organisms are injected, causing the body to produce memory cells.

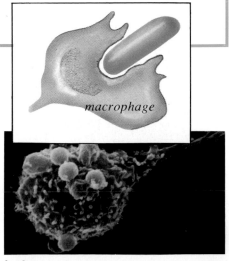

leukocyte devouring bacteria

Breathing

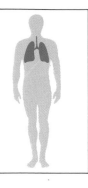

Respiration means *breathing*. We breathe because the cells in the body need oxygen to burn up food and because the burning process produces carbon dioxide which the body must get rid of. The respiratory system consists of air passages and lungs.

The air we breathe

About 20% of air is oxygen (O_2). The remainder is mainly nitrogen (N_2), plus a little water vapour (H_2O), carbon dioxide (CO_2) and a few other gases. The air we breathe out is broadly similar to what we breathe in but contains somewhat less oxygen (about 16%) and somewhat more water vapour and carbon dioxide.

The air passages

When we breathe in, the air passes through the air passages (the nose, the throat, the paranasal sinuses, the larynx, the windpipe or trachea and the bronchi) to the lungs.

The air we breathe in is warmed up in the *nose* and cleansed as it is filtered through the *nasal hairs*. When we breathe out, the air in the nose is cooled down, so that the body does not lose too much heat. Air inhaled through the *mouth* is not warmed up, cleansed or filtered so well.

The air passes through the *throat (pharynx)* and the *voice box (larynx)*

The *windpipe (trachea)* has a mucous membrane lined with small, thin hairs (*cilia*). Here the air is further cleansed, warmed and moistened before reaching the lungs. The windpipe is strengthened with rings of cartilage.

See also:
The blood (22, 23)
The circulation (18, 19)
Speech and respiration (28, 29)

The *paranasal sinuses* are air-filled, hollow cavities in the facial skull. They act as a *sounding box* when we talk.

main bronc

The lungs

From the windpipe the air passes through two *main branch tubes (main bronchi)* and on to the two lungs *(pulmones)*. The right lung is divided into three *pulmonary lobes* and the left lung into two.

Between the lungs and the thorax there is a narrow space filled with fluid, the *pleural cavity*. Both the outer surface of the lungs and the inside of the thorax are covered with a thin, smooth membrane – the *pulmonary membrane (pleura)*. This membrane, together with the fluid in the pleural cavity, ensures that as we breathe, the lungs slide smoothly against the rib cage.

The main bronchi are divided into a large number of smaller *branched air tubes (bronchioles)*. There, the air is inhaled and expelled by several hundreds of millions of bubble-like *airsacs (alveoli)*. Because the alveoli are so small, and there are so many of them, they form a total surface area of 60-100 m^2.

Around the alveoli there is a network of capillary tubes.

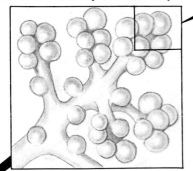

Gas exchange

The walls of the alveoli and capillary tubes are very thin. Gases can easily be exchanged through these walls between the air and the blood. In this way oxygen from the air can enter the alveoli and then pass into the blood in the capillaries. Here it bonds itself to the haemoglobin in the red blood cells.

This bonding ensures that the haemoglobin acquires a fresh red colour. Carbon dioxide passes from the blood to the air in the alveoli. In this way, the blood is provided with oxygen and gets rid of carbon dioxide.

$O_2 \rightarrow O_2$

$CO_2 \leftarrow CO_2$

From the lungs the blood is transported by the lesser circulation to the heart and is then pumped throughout the body in the greater circulation. In the capillaries about one quarter of the oxygen is transferred to the tissue cells. The remainder stays in the blood as a reserve.

Carbon dioxide gas is formed as a waste product from the combustion process within the cells. It is transferred from the tissue to the bloodstream. From the capillaries the blood is transported in the greater circulation to the heart and then is pumped to the lungs via the lesser circulation.

Diseases

Asthma is a disease where the bronchial tubes suddenly close up, firstly because the muscles contract and secondly due to the swelling of the mucous membrane. It thus becomes difficult to breathe out. Asthma may have several possible causes: allergy and psychological stress are two of them.

The mucous membranes in the air passages react to infections by swelling up and increasing the production of mucus. Such infections are called bronchial infections. The infections are given names according to the place where they are found, such as *paranasal sinusitis, bronchitis* (in the bronchial tubes) and *inflammation of the lungs* (pneumonia). We normally call infections in the other air passages common colds. As a rule a virus is responsible for the infection, but sinusitis and pneumonia can regularly be caused and aggravated by bacteria. Some people have a violent reaction to certain substances, so that the mucous membrane swells up as it does with a bronchial infection. This is a form of allergy.

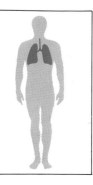

Speech and breathing

When we breathe we suck air into the lungs and expel it out again. When we tighten the vocal cords as we breathe out, we can make them vibrate and make a noise. That is a precondition of being able to speak.

Breathing

Breathing (*respiration*) means sucking air into the lungs and then expelling it. At rest we breathe in (*inhale*) and out (*exhale*) about 15 times a minute. Normally we breathe approximately half a litre of air in and out. When we have exhaled, there are still about 2 litres of air left in the lungs to prevent them from collapsing. When we exert ourselves, we can also exhale some of this air.

When the body is active, it also needs more oxygen. We provide for this by breathing in and out more quickly and more deeply. Then we use muscles which are attached to the chest cavity (thorax) in conjunction with the *diaphragm*. These help with the quicker, more frequent expansion and contraction of the thorax, so that we breathe faster. An excess of carbon dioxide (CO_2) in the blood stimulates the breathing centre in the brain. This sends impulses to the respiratory muscles.

See also:
The respiratory system
(26, 27)

For more information about the speech centre, see:
The brain (cerebral cortex) (35)

See also:
The brain stem (33)

Around the lungs lie the thorax and the *midriff* (*diaphragm*). As well as the thorax, the lungs and diaphragm are covered with membranes (*pleural sacs*), between which is a very narrow, fluid-filled space *(the pleural cavity).*
We are able to use the muscles in the diaphragm to draw this downwards. Although the lungs do not rest directly on the diaphragm, a thin layer of fluid in the pleural cavity means that the base of the lungs is nevertheless bonded to the diaphragm. Thus the lungs are pulled with it when the diaphragm moves downwards. This makes the lungs expand and air is drawn into them, just as air is drawn into a pair of bellows when these are opened out.

When anything irritates the mucous membrane in the windpipe or the lungs, we *cough*. Coughing is a reflex action, whereby we first breathe in quickly, close the larynx with the vocal cords, tighten the muscles in the thorax so that a high pressure occurs and opens the larynx by relaxing the vocal cords. The air rushes up through the air passages at a speed of 150 km/hour. Frequently the air will carry with it whatever has irritated the mucous membrane, so that we 'cough it up'. *Sneezing* makes one think of coughing but with sneezing the air passes through the nose in order to clear it of whatever irritated the mucous membrane.

Speech

In the larynx there are two folds of connective tissue - the *vocal cords*. When we tighten the vocal cords and breathe out, they begin to vibrate so that a sound is formed. If we tighten the vocal cords even more, then the sound becomes higher; by relaxing them, the sound gets lower.

We can also change the sound made by the vocal cords by moving the mouth, tongue and lips. We speak by combining the various sounds.

Although we are born with vocal cords and larynx, we have to learn to talk by imitating the sounds we hear from other people. Speech is controlled by centres in the dominant half of the brain. With most right-handed people, this is in the left hemisphere.

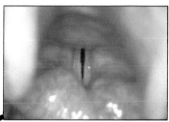

tightened vocal cords (when speaking)

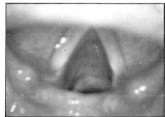

relaxed vocal cords

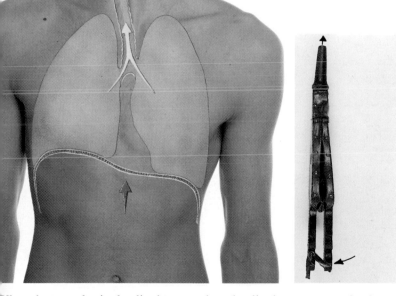

When the muscles in the diaphragm relax, the diaphragm moves back up again and the air is expelled out of the lungs. Breathing is controlled by a nerve centre in the brain. From there, signals are sent to the muscles, so that we breathe in and out without having to think about it. Respiration is controlled by, among other things, the need for oxygen.

Diseases

Smoking damages the air passages in many ways. Nicotine paralyzes the cilia, so that the air we inhale is not cleaned so well. Nicotine is also the cause of blood vessels contracting, reducing the flow of blood to the body. Smoking irritates the mucous membranes, so that they frequently become chronically inflamed (chronic bronchitis). This produces a chronic *smoker's cough*. The tar in the smoke remains behind in the lungs, so that they become completely black. There is also a clear indication that the chances of contracting cancer are increased by smoking.

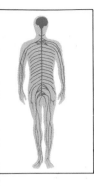

Sensory nerve endings and nerve cells

The nervous system consists of sensory nerve endings and nerve cells (*neurons*). The sensory nerve endings produce *electric impulses* (*signals*) when they are stimulated and the neurons conduct the electric impulses further. Offshoots from the neurons, the nerve fibres, act as electrical conductors, taking information from the senses to the central nervous system and from the central nervous system outwards through the body.

The senses

The sensory organs consist of specialized sensory cells. Each sensory cell reacts only to one kind of stimulus. Some sensory cells react to light, others to noise, yet others to pain, touch, etc. When sensory cells are exposed to their precise type of stimulus, they produce electric impulses.

Information about visual perceptions are conducted to one part of the brain, whilst information about sounds goes to another part, and so on. In this way, the brain can distinguish between the different stimuli. Vision, hearing and balance are important senses, and they are therefore dealt with as separate subjects.

The sensory cells at the top of the nose perceive chemical matter in the air as smells. We are able to distinguish several dozen smells.

Feeling in the skin is called *superficial sensitivity*. Tiny sensory organs (*receptors*) register touch, pressure, heat, cold and pain.

Depth sensitivity is the feeling within the body. Tiny receptors perceive the position and movement of joints and muscles and pain in the intestines.

For more about the senses, see:
Hearing and balance (38, 39)
Sight (40,41)

For more about the nervous system, see:
The central nervous system (32, 33)
The brain (34, 35)
The peripheral nervous system (36, 37)

See also:
Tissue (15)

Taste buds on the tongue can distinguish between salty, sweet, sou and bitter. The 'taste' of food and drink is composed of these fou tastes and from impulses from the sense of smell.

The nerve cells

Each *nerve cell* (*neuron*) consists of a cell body and some offshoots, nerve fibres. Nerve fibres which conduct electrical impulses to the cell body are called *dendrites*. Most neurons have many dendrites.

Each neuron always has one *axon*. That is a nerve fibre which conveys electrical impulses from the cell body. Axons can be very long (more than one metre).

Each nerve fibre has an insulating white sleeve, called the *myelin sheath*. These fibres are able to conduct electrical impulses very quickly (more than 100 metres per second). They are important because they inform the central nervous system extremely rapidly about any harmful stimuli (for example, when we burn ourselves) and send information to the muscles so that they react quickly (by moving the hand away, for example).

The majority of nerve fibres lack the myelin sheath. They conduct the electrical impulses more slowly. Nerve fibres from the sensory organs to the central nervous system and from the central nervous system into the body are bundled into cables which we call nerves. One nerve may consist of dozens of nerve fibres.

cell body

dendrite

axon with myelin sheath

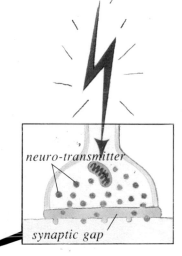

neuro-transmitter

synaptic gap

The axon never lies fully against the cell which it must provide with information but is separated from it by a thin, fluid-filled gap. The electrical impulse is conducted from one cell to the next via a chemical connecting point which is called a *synapse*. When an electrical impulse arrives, the end of the axon excretes a chemical (*neuro-transmitter*) into the gap. This matter stimulates the other cell to make a new electrical impulse.

axon without myelin sheath

Glia cells are special connective tissue cells which support and protect the nerve cells.

Many nerve cells form a synapse with the cell body and the dendrites of other nerve cells. One cell may be in contact with many other nerve cells. Some nerve cells form a synapse with other kinds of cells, for example muscle cells. It is the physical and chemical processes occurring in the muscular tissue which, after transmission over the synapses, lead to the contraction of the muscle.

31

The central nervous system

The central nervous system (CNS) is the command centre which controls the body. This consists of the brain and the spinal cord. The brain is divided into the cerebrum and the cerebellum. The central nervous system is protected by the cerebral membrane (*meninges*) and fluid (*cerebro-spinal fluid*).

For more information about the brain hemispheres and their tasks, see:
The cerebrum (34, 35)

See also:
Sensory nerve endings and nerve cells (30, 31)
The peripheral nervous system (36, 37)

Grey and white matter

The areas in the central nervous system which have many neurons and nerve fibres without myelin sheathing are grey (*grey matter*). Areas with many nerve fibres with myelin sheathing are white (*white matter*).

The cerebellum

The *cerebellum* controls and supervises movement and balance. It receives information through the nerve fibres about the position and movements of the body. Voluntarily controlled movements start from the movement centre in the cerebrum. The cerebellum coordinates the muscles; the movements are even, rhythmical and supple and are performed with the right amount of effort at the right time.

The cerebral membranes

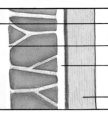

spinal cord
pia mater
arachnoid mater
dura mater

The central nervous system is covered by three protective membranes (*meninges*). The *hard membrane* (*dura mater*) is the outermost and hardest. This covers the inside of the skull and the spinal canal. The *soft membrane* (*pia mater*) is the innermost of the cerebral membranes. This lies directly over the surface of the brain and against the surface of the spinal cord. The *arachnoid mater* is the middle membrane. This is like a cobweb between the hard and soft membranes.

The cerebro-spinal fluid

This fluid is produced in the walls of four hollow cavities (ventricles) in the brain. The fluid flows through narrow channels between the ventricles and from there into the space between the arachnoid and inner membranes. It surrounds the whole central nervous system and protects it against knocks. From the space under the arachnoid membrane the fluid enters the blood. An equivalent amount of fluid is carried away as is produced so that pressure remains constant.

Underneath, at the top of the spinal cord between the cerebral membranes, there is a large cavity filled with fluid. Doctors are able to remove some of the fluid from this point for examination. The manner in which this is done is called a *cisternal puncture*. Such an examination is done to ascertain whether there are any white blood corpuscles (in the case of meningitis) or red blood corpuscles (in simple cerebral haemorrhages) or even foreign proteins in the fluid.

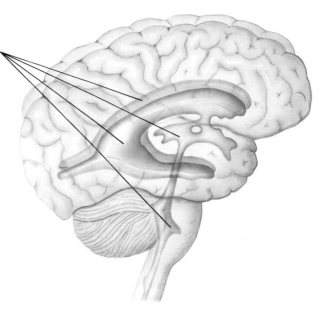

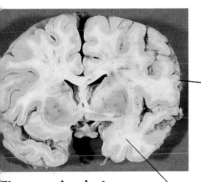

The cerebral stem

The *cerebral stem* forms the connection between the cerebrum,

the cerebellum and the spinal cord. The cerebral stem contains groups of nerve cells which control, among other things, breathing and blood circulation. The neurons receive and send impulses via the cerebral nerves. At the top of the cerebral stem is a control point (*thalamus*) which among other things determines which sensory impulses must be passed to the brain and which are unimportant and can be withheld.

The lowest part of the cerebral stem adjacent to the spinal cord is called the *medulla oblongata*. Here is based the breathing centre among others.

The cerebrum

The *cerebrum* weighs about 1300 grams. It consists of two halves (*hemispheres*).

The *cerebral cortex*, the outermost layer of the cerebrum, contains millions of nerve cells. Consciousness is located in the cerebrum and the sensory perceptions are processed as the nerve impulses reach certain areas of the cerebral cortex. White matter in the cerebrum consists for the most part of nerve fibres to and from the cerebral cortex.

The spinal cord

On the outside of the *spinal cord* (*medulla spinalis*) there are nerve tracts to and from the brain; inside these are neurons. Most of them are in contact with other nerve cells; some have branches going to the body via the nerve tracts. In the topmost and lowest parts of the spinal cord there are a great many nerve cells, because the muscles in the arms and legs are controlled from there.

Diseases

A *stroke* is caused by a blood clot in the blood vessels to the brain (a clot occurs because the blood in the blood vessels congeals so that the flow of blood is blocked). Blood clots in the right half of the brain can lead to paralysis of the left side of the body. In the left half of the brain they can lead to paralysis of the right half of the body and then cause speech impediments. A large blood clot can be fatal.

A *cerebral haemorrhage* (bleeding of the blood vessels in the brain) can give the same symptoms but occurs less frequently.

Toxic substances (including alcohol and drugs) can seriously damage the brain and nerve cells.

Inflammation of the cerebral membrane (*meningitis*) is caused by bacteria or viruses. *Cerebro-spinal meningitis*, one form of inflammation of the cerebral membrane, is caused by special bacteria called meningococci. Meningitis is a serious disease and can be fatal.

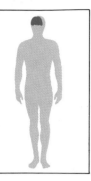

The brain

The brain (cerebral cortex) is superior to the rest of the central nervous system; it processes sensory impressions and controls the voluntary movements. Coordination between the neurons in the cerebral cortex is responsible for speech, memory, intelligence, thinking and the various moods.

The tasks of the cerebral cortex

Despite the fact that we know what tasks the various small areas in the brain have, we know very little about the large areas in between them. Probably consciousness, the ability to think, touch and intelligence come from these areas. All these functions are extremely complicated and demand the coordination of millions of neurons in various parts of the brain. However, nobody has as yet been able to discover precisely how such characteristics occur in the brain.

Nor do we know precisely how the various impressions are stored in the brain. It is known however that memory exists because new contacts occur between neurons in the cerebral cortex. The memory is stored in various parts of the brain. We think that short-term memory (the memory of things which do not have to be remembered for long) is in a different place from the memory of events which influence our existence and thus have to be remembered 'for ever'.

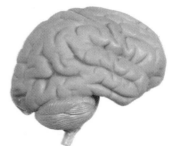

The visual cortex

We become aware of what we see in the cerebral cortex at the back of the occipital lobe. It is not true that we only perceive light and colours. In the areas around the visual cortex, we link the images together and in this way we are able to evaluate what we see and recognize people, animals and objects.

For more about the brain, see:
The central nervous system (32, 33)

See also:
Sensory nerve endings and nerve cells (30, 31)
The peripheral nervous system (36, 37)
Hearing and balance (38, 39)
Sight (40, 41)

central groove

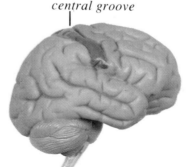

The touch cortex

We are conscious of what we feel, for example touch and pain, in a small area which lies directly behind the motor cortex. This is the foremost part of the parietal lobe. The groove which separates the motor cortex from the sensory (touch) cortex is called the central groove. In order to ensure that the body reacts quickly to touch impressions, it is important for the motor cortex to be close to the touch cortex with good connections between them.

central groove

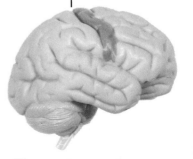

The motor cortex

Voluntary muscular movements are controlled from a small, well delineated area in the cerebral cortex. This is the rearmost part of the frontal lobe. Coordination of the body's movements is, however, controlled by the cerebellum.

The two halves of the brain (*hemispheres*) resemble mirror images of each other. The right hemisphere controls the left half of the body, and the left hemisphere controls the right. However, for individual characteristics the two hemispheres have quite different tasks. In the case of 97% of people, the left hemisphere is responsible for speech and numeracy (dominant hemisphere). For the other 3% it is the right hemisphere. The dominant hemisphere works like an accountant, whilst the other half is more artistic. When we sing, it is the dominant hemisphere which reads the text, whilst the non-dominant hemisphere is aware of the melody.

Sleep

During sleep, awareness is switched off and the brain rests. Nevertheless, there is still some activity in the brain. During dreaming, for example, there is a lot of activity in the cerebral cortex. We are completely dependent on regular sleep. After a few days without sleep, we become physically and mentally exhausted. Some people receive *visions* and are plagued by other problems. However, we still know very little about our dependency on a regular sleep pattern.

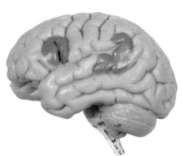

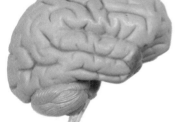

right hemisphere   *left hemisphere*

The centres of hearing and speech

The hearing and speech functions are in different areas of the cerebral cortex. A small area in the temporal lobe perceives sound. From there, signals are sent to a speech centre, which interprets words and sound stimuli. An internal image is formed of what we hear. When we ourselves speak, the brain first forms an internal image of what we are going to talk about. Then, signals are sent to another speech centre (in the frontal lobe) which looks for words to 'express' these images. When the correct words have been found, the information is then channelled to the motor cortex. There the brain determines which muscles in the larynx must be coordinated in order to produce the correct sounds.

The centre of smell

Although we have a much poorer sense of smell than most animals, we are still able to distinguish between several dozen smells. The smell is perceived in the temporal lobe (the lobe of the brain 'under' the temporal bone).

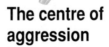

The centre of aggression

Just under the cortex in the temporal lobe there is a collection of nerve cells which govern temper and anger amongst other things.

The peripheral nervous system

The nerves and their branches form the peripheral nervous system. The task of the nerves is to pass on messages from the body to the central nervous system and from the central nervous system to the organs of the body. There are two main types of nerves: cranial nerves and spinal nerves. We make a distinction between the part of the nervous system which is consciously controlled (*somatic*) and that part which is not consciously controlled (*autonomic*).

The cranial nerves and the spinal nerves

The *cranial nerves* run in pairs from the cerebral cortex to their own half of the brain (12 pairs altogether). The optic nerve, the olfactory nerve, the auditory nerve and nerve of balance are examples of cranial nerves. The *vagus nerve* is the longest of the cranial nerves. This controls the activities of the heart and most of the digestive tract.

The *spinal nerves* emerge from the spinal cord in pairs (31 pairs in total): the cervical nerves, the thoracic nerves and lumbar nerves, the sacral nerves and coccygeal nerves. The spinal nerves branch off throughout the whole body. A spinal nerve and the area which it covers can be rendered numb by injecting an anaesthetizing fluid into the outside of the hard membrane, where the nerve emerges from the spinal cord (*epidural anaesthetic*).

Reflexes

Impulses from the body reach the spinal cord via the spinal nerves. The spinal cord sends these messages to the cerebral cortex. There they are put in order and it is determined how the body is to react. The messages back to the body from the brain are sent to the spinal cord and channelled further by the spinal nerves. Sometimes a fast reaction is required, for example in order to protect the body from injury. Then the decisions are taken automatically in the spinal cord (or in the brain stem) and fresh nerve impulses are directed to the body so fast that the cerebral cortex cannot influence them. These we call *reflexes*.

For more about the nervous system, see:
Sensory nerve endings and nerve cells (30, 31)
The central nervous system (32, 33)
The cerebral cortex (34, 35)
For the names of the peripheral nerves, see:
Anatomical terms (89)

cranial nerves

spinal nerves

Reflex actions are born in us, but some reflexes disappear after a while. One such reflex is a baby's *sucking reflex*, which ensures that the baby is able to be breast-fed without first having to learn to suck.

Reflex actions occur independently of conscious control, so that the same stimulus always leads to the same reaction. A number of reflexes are coupled to the *voluntary (somatic)* part of the nervous system and act on the muscles. A number of others are connected to the *involuntary (autonomic)* part of the nervous system. These reflexes are important for controlling the activities of the internal organs.

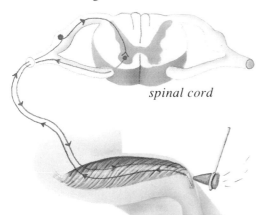

spinal cord

The *knee jerk reflex* is another example of a reflex action. When we hit the tendon below the kneecap, stretching occurs of the extensor muscle at the front of the thigh. Sensory cells in the muscle perceive this stretching and send a message to the spinal column. In order to counteract this stretching, the spinal column sends an instruction that the extensor muscle must relax. Simultaneously a signal is sent to the cerebral cortex, so that it knows what is happening.
With pain, for example when we burn our hands, we pull the arms back, because all the flexor muscles contract and the hands are withdrawn instantly by reflex action.

The voluntary nervous system
Normally the activity in the skeletal muscles is under conscious control. The nerves of these muscles are part of the voluntary (somatic) nervous system.

The autonomic nervous system
Smooth muscular tissue, cardiac tissue and *glands* are controlled by reflexes in the nerves which are part of the autonomic (involuntary) nervous system. Many of these reflexes are of vital importance. The autonomic nervous system consists of two parts, the *sympathetic* and the *parasympathetic* parts. Sympathetic and parasympathetic nerves have opposite effects: one increases the activity within an organ and the other slows it down. The nerve fibres belonging to the autonomic nervous system run from the brain stem and the spinal cord to the various organs of the body.

The *sympathetic nervous system* is a crisis system. With worry, anger and physical tension, the pulse, blood pressure and perspiration increase and the pupils enlarge. These changes place the body in a better condition to fight off the crisis.

The *parasympathetic nervous system* works especially when the body is at rest. It lowers the pulse rate and the blood pressure and increases intestinal activity, so that food is digested. The vagus nerve is the most important parasympathetic nerve.

Sensory and motor nerve fibres
Nerve fibres which conduct touch impulses from the sensory organs to the central nervous system are called *sensory nerve fibres*. Nerve fibres which take messages from the central nervous system to the muscles that they have to move are called *motor nerve fibres*. Most nerves have both motor and sensory fibres. The *sciatic nerve* stems from some of the lumbar and sacral nerves. It is the largest nerve in the body and carries sensory fibres from the legs and motor fibres to the legs.

Hearing and balance

Both the hearing organ and the balance organ are located in the ear. Sounds spread as vibrations through the air (sound waves). The ear can perceive these vibrations. Sensory cells in the inner ear convert the vibrations into electrical impulses, which are sent to the brain. In the inner ear are also some tiny sensory organs which continuously record the position and movements of the head.

The outer ear

The outer ear acts as an antenna which picks up sound waves. The outer ear consists mainly of cartilage.

The auditory canal conducts the vibrations further.

The middle ear

The middle ear is the space behind the eardrum. Three tiny bones called ossicles in the middle ear (*hammer, anvil* and *stirrup* or *malleus, incus* and *stapes*) transmit the vibrations from the eardrum to the *oval window* in the inner ear.

The inner ear

The *cochlea* is part of the inner ear, where the specialized hearing cells are found.

Fluid and membranes are made to vibrate by the ossicles. In the lowest part of the cochlea the membranes are narrow. They are made to vibrate by high sounds (fast vibrations). Membranes close to the top of the cochlea are broader. They are stimulated by lower sounds. When the membranes vibrate against each other, the sensory cells create electrical impulses on one of the membranes. The nervous impulses are then sent to the special hearing centre in the cerebral cortex.

The *eardrum* is made to vibrate by the air in the auditory canal.

The *Eustachian tube* connects the middle ear to the throat. It opens out behind the nose and ensures that pressure is equal on both sides of the eardrum. When pressure is unequal, hearing deteriorates and we feel as if the ear is blocked. This can happen with a cold, when the mucous membrane behind the nose swells and cuts off the airflow in the tube.

For more information about sensory cells, see: Sensory nerve endings and nerve cells (30, 31)

For more about the hearing centre, see: The cerebral cortex (35)

Diseases

The causes of hearing loss may be in all parts of the ear. The accumulation of ear wax in the auditory canal, punctures in the eardrum, calcification between the ossicles and damage to the sensory cells in the inner ear are the usual causes of hearing loss. Noise can lead to the destruction of sensory cells in the cochlea. *Inflammation of the middle ear* (otitis media) is a very common ear complaint.

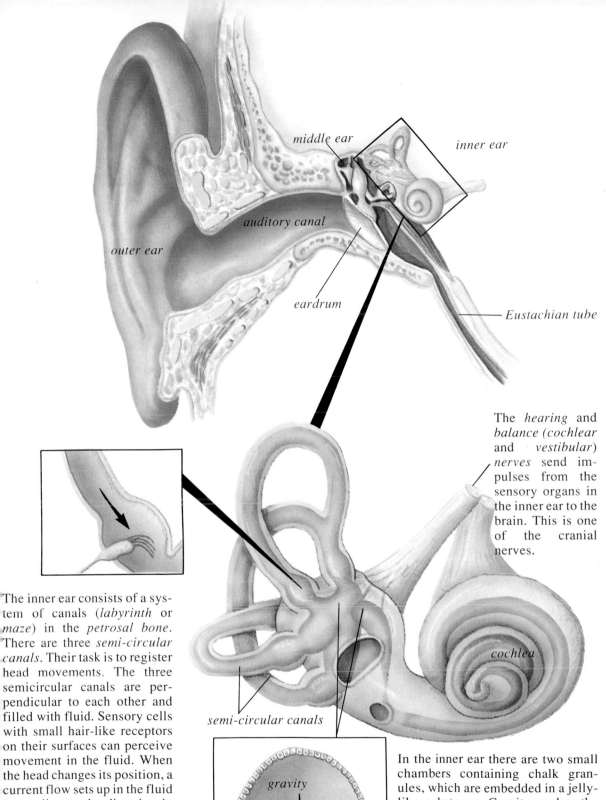

middle ear

inner ear

auditory canal

outer ear

eardrum

Eustachian tube

The *hearing* and *balance (cochlear* and *vestibular) nerves* send impulses from the sensory organs in the inner ear to the brain. This is one of the cranial nerves.

The inner ear consists of a system of canals (*labyrinth* or *maze*) in the *petrosal bone*. There are three *semi-circular canals*. Their task is to register head movements. The three semicircular canals are perpendicular to each other and filled with fluid. Sensory cells with small hair-like receptors on their surfaces can perceive movement in the fluid. When the head changes its position, a current flow sets up in the fluid (according to the direction in which the movement occurs). The sensory cells then create nervous impulses which are sent to the brain, so that the brain is kept informed of the movements and position of the head.

cochlea

semi-circular canals

gravity

In the inner ear there are two small chambers containing chalk granules, which are embedded in a jelly-like substance. Gravity pushes the crystals to the lowest part of the chamber. Cilia-covered sensory cells continuously monitor the position of the crystals and direct nervous impulses to the brain with information about the position of the head.

39

Sight

Light passes through the air in beams. Light beams pass through the cornea and penetrate deeper into the eye via the pupil. The light beams are focused on the retina at the back of the eye, where sensory cells perceive the light and create nervous impulses. The nervous impulses are sent to the brain, which interprets what is seen.

The structure of the eye

The *cornea* has no blood vessels and is crystal clear. It forms a window at the front of the eye, so that the light can shine through.

The *anterior chamber* behind the cornea contains a clear fluid called the *aqueous humor* which has to supply the cornea.

The *iris* contains a brown colouring (*pigment*) to prevent the penetration of light. Therefore, all light has to pass through the *pupil*, which forms an aperture within the centre of the iris. The quantity of pigment in the iris determines the colour of the eye.
Muscles in the iris control the size of the pupil. When the light is bright, the pupil becomes smaller in order to let in less light; as it gets dark, the pupil enlarges to let in more light.

For more about sensory cells, see:
Sensory nerve endings and nerve cells (30, 31)

For more about the visual cortex, see:
The cerebral cortex (34)

The *lens* is suspended by means of an orbicular muscle. When this muscle contracts, the lens becomes round. When the muscle relaxes the lens becomes oval (flatter).

Six small muscles on the outside edge of each eye are able to move it in all directions. The eye muscles originate in the wall of the eye socket and are fixed to the sclera.

The *blind spot* is a small spot where the optic nerve leaves the eye. There are no sensory cells at that point.

The *choroid membrane* distributes the blood supply to most parts of the eye.

The *sclera* forms the outside of the eye. This is thick and leathery. The visible part of the sclera is covered with a thin membrane called the *conjunctiva*. When we get an inflammation or any dirt in our eyes, the blood vessels in the conjunctiva dilate so that the eye becomes red.

The *vitreous humor* is a jelly which fills most of the eye. The jelly is clear and does not hinder the path of the light to the retina.

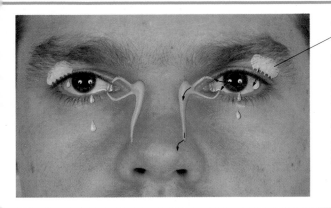

The *tear gland* (*lacrimal gland*) secretes tears. These are spread over the eye as we blink. Tears cleanse the eyes and prevent them from drying out. Tears are conducted to the *lacrimal sacs* and from there to the nose. Only when we produce a lot of tears (as when we cry) do the tears run down the cheeks.

The *eyebrows* and *eyelids* with the *eyelashes* protect the eyes from foreign bodies and other matter.

The *retina* contains sensory cells which perceive the light and send nervous impulses to the brain. The extremely light-sensitive cells, which can however only distinguish between light and dark, are called *rods*. The sensory cells which can distinguish colours are called *cones*.

Most of the cones are to be found in the *macula* on the retina. This is where we see the sharpest, best image in daylight.

The *optic nerve* conducts the nervous impulses from the sensory cells to the brain.

Diseases

Cataracts occur frequently with age. With this the lens becomes cloudy, so that the light rays can only penetrate with difficulty and vision becomes poor.

Glaucoma occurs because pressure builds up on the fluid in the eye. When the pressure gets very high, the sensory cells are damaged and the eye becomes blind.

Night blindness can be caused by a lack of Vitamin A. Rods in the retina need Vitamin A in order to be able to function normally.

The light path

When light rays hit the cornea, they are refracted (bent) towards the middle of the eye. In the lens the light is further refracted. The refraction of light in the cornea is constant, but may be varied in the lens, so that we can see sharply at long and short distances (*accommodation*). When we focus close up, the lens is round so that the light is greatly refracted. When we focus in the distance, the lens is oval, so that the light is less refracted. In both cases, the light rays converge on the retina in and around the macula. The image that we see is upside down on the retina. The brain turns the picture round so that we perceive the image 'normally'. In the case of short-sightedness (*myopia*) the light is refracted too much so that the light rays converge in front of the retina. In the case of long sight (*presbyopia*) the light is insufficiently refracted so that it converges beyond the retina.

As we age the lens gets less supple so that the eye can no longer refract the light adequately to be able to see sharply at short distances (ageing long-sightedness or hyperopia).

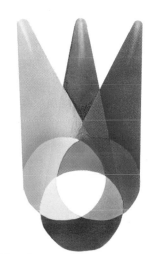

Seeing colours

We have three types of cones in the retina. One type perceives red light, one blue and one green. By blending red, blue and green, we are able to identify all the colours. (Blending divergent light, however, gives different colours compared with blending similar colours.) When the nervous impulses from the three types of cones are perceived in the brain, it is able to distinguish between all the shades of colour.

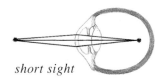

short sight

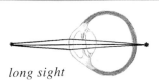

long sight

41

Hormones

Hormones are stimulants which are involved in the control of the body's functions. They are produced by endocrine glands and discharged into the blood. The blood carries the hormones round the whole body, where they influence the cells. Each hormone has its own specific tasks.

For more information about the sex hormones, see:
Puberty (64)
Menstruation (70, 71)
Pregnancy and birth (76, 77)

How hormones work

The *internal secretion* or *endocrine* (= *hormone producing*) *glands* have no ducts, unlike the *exocrine* glands. They pass the hormones directly into the blood, so that they are quickly pumped around the body with the blood.

The quantity of specific hormones in the blood is subject to large fluctuations. The quantity of female sex hormones, for example, varies during the different phases of the menstrual cycle. The quantity of other hormones present in the blood is virtually constant. The amount of hormones is controlled by *feedback*: when there is too little of any hormone, the secretion of that hormone in the body is increased. When there is too much of a certain hormone, secretion is reduced.

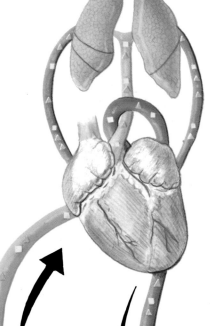

hormones in the blood

cell with receptor

In the body the hormones will fasten on to special receivers (*receptors*) on the cells.

gland

We can compare hormones to keys and receptors to keyholes. Hormones can only act on cells with receptors which they fit. When the hormone latches itself on to the receptors, changes take place within the cell. When, for example, insulin attaches itself to a cell's receptors, then the cell starts to absorb glucose (blood sugar) from the blood.

Endocrine glands

The *thyroid gland* produces the hormone *thyroxine*, which controls the combustion of nutrients (*metabolism*).

The *parathyroid glands* lie behind the thyroid gland. Their hormone (*parathormone*) is concerned with controlling the amount of calcium in the body.

The *adrenal cortex* secretes special kinds of hormones; these are called *steroids*. *Aldosterone* controls the salt excretion in the kidneys. *Cortisone* influences metabolism and slows down inflammatory reaction. The sex hormones are also steroids. The adrenal cortex produces small amounts of male and female sex hormones.

The *adrenal medulla* secretes *adrenaline* and *noradrenaline*. They work together with the sympathetic nervous system and raise the pulse rate and blood pressure.

The endocrine cells (Islets of Langerhans) in the *pancreas* produce *insulin* and *glucagon*. Insulin lowers the sugar content of the blood and stores it in the cells. Glucagon works in reverse. The interplay between the two controls the sugar content in the blood.

The *pituitary gland*, on the underside of the brain, is an important endocrine gland. It secretes hormones concerned with urine production, skeletal growth, the activity of the sex glands and the mammary glands in women. Some hormones from the pituitary control the secretion of hormones by the thyroid gland, the adrenal cortex and the sex glands.

The *ovaries* in women and the *testicles* in men produce *sex hormones*.

Diseases

Diabetes (*diabetes mellitus*) occurs when the pancreas produces too little insulin. Too much sugar passes into the blood and sugar appears in the urine. Diabetic patients have to inject themselves with insulin in order to control the level of sugar in the blood.

In the case of *increased metabolism* the cells burn up more nutrients, so that the body's weight falls, even if it is eating more.

This is caused by too much thyroxine hormone secretion. In many patients with increased metabolism the thyroid gland enlarges. This we call *struma* (*goitre*). Struma can occur when the thyroid gland is not in a condition to produce thyroxine. Stimulation by the pituitary then leads to growth of the thyroid gland.

Digestion

The cells in the body need nutrients in order to be able to live. The digestion breaks down the food that we eat, so that the nutrients from the intestines can be absorbed and passed on to the blood. The blood then carries the nutrients to the cells throughout the body. The materials in the food that the body cannot use are passed through the intestines and evacuated as faeces.

Diet

The cells in the body need energy in order to live. Proteins, fat and carbohydrates are the most important nutrients. The cells convert them into heat or other energy or use them as building materials. Besides these nutrients there are yet other materials which the body always needs in small amounts. Amongst these are salts and minerals (sodium, potassium, chloride, iron, magnesium, sulphur and phosphor) and vitamins (A, B, C, D, E and K). A varied diet as a rule provides all these materials for our requirements.

The energy requirement varies

The energy intake requirement depends on body weight and muscular activity. Too high an energy intake (which means that we eat more than we need leads to the nutrients being converted into fats which are absorbed into the body's fatty tissue. Then we get too heavy and that is unhealthy. If the energy intake is too low, this leads to weight loss and then to malnutrition. Obviously that too is harmful.

For details about the organs in the digestive system, see:
The digestive tract
- upper section (46, 47)
The digestive tract
- middle section (48, 49)
The digestive tract
- lower section (50, 51)

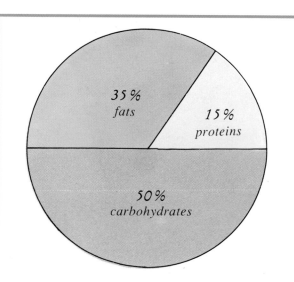

In a well constituted diet, the proteins must cover about 15% of the energy supply, fats less than 35% and carbohydrates the rest.
In Europe, most people eat too much fat and carbohydrates and not enough dietary fibre.

The digestive tract

The gullet, stomach and intestines form a coherent whole (the digestive tract or alimentary canal). The inside of this canal is lined with *mucous membrane*. This consists of an epithelium lining over a thin layer of connective tissue. The walls of the tract consist of smooth muscular tissue. The canal is in many places covered with *peritoneum*, for example on the outside of the stomach and on the major part of the intestines.

The *teeth* masticate the food that we eat to break it into smaller pieces.

Saliva from the *salivary glands* moistens the masticated food and helps to break down the *starch*.

The *oesophagus* carries the food to the stomach.

The *stomach* kneads the food and mixes it with *digestive juice*, which plays a role in breaking down the nutrients.

One of the many tasks of the *liver* is to make gall. Gall is stored in the *gall bladder* and excreted into the duodenum as required. The gall ensures that the fat is broken down into small drops.

The *duodenum* is the uppermost part of the *small intestine*. Here the food is further broken down. The nutrients are absorbed out of the small intestine by the blood.

Pancreatic juice from the *pancreas* is excreted into the duodenum. The pancreatic juice breaks down the nutrients so that they can be absorbed.

Water and electrolytes are absorbed into the blood from the *small intestine* and the *colon*.

The *rectum* stores up the faeces and evacuates the faeces.

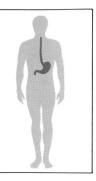

The digestive tract - upper section

In the upper section of the digestive tract or alimentary canal the first digestive processing of the food takes place. Only when the food has been kneaded together with gastric juice to form a viscous mass is it conveyed further out of the stomach.

The mouth

The *tongue* has a rough surface. Between each of the protuberances are taste buds (*taste papillae*), which differentiate between salty, sweet, sour and bitter tastes. Inside the tongue there are striated muscles.

surface of tongue (greatly enlarged)

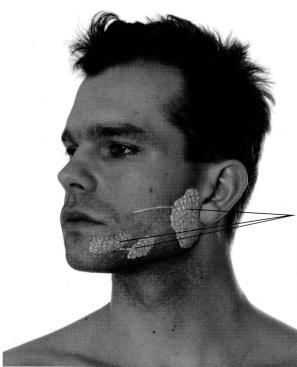

Three pairs of large *salivary glands*, including the *parotid glands* in front of the ear, which swell in the event of mumps) and many small salivary glands, secrete saliva into the oral cavity. The saliva moistens the food and lubricates the throat so that we can swallow more easily. Enzymes in the saliva help to break down starch in the food.

For more information about the digestive system, see:
The digestive system (44, 45)
Digestive tract - middle-section (48, 49)
Digestive tract - lower section (50, 51)

The first teeth, the *milk or deciduous teeth*, consist of 20 teeth. From the fifth or sixth year of age these are gradually replaced by the *permanent or second teeth*, which consist of 32 teeth.

The *incisors* and *canines* are sharp teeth so they are able to bite and cut up the food.

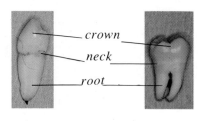

crown

neck

root

The *molars* are rough so they are able to grind and chew the food.

46

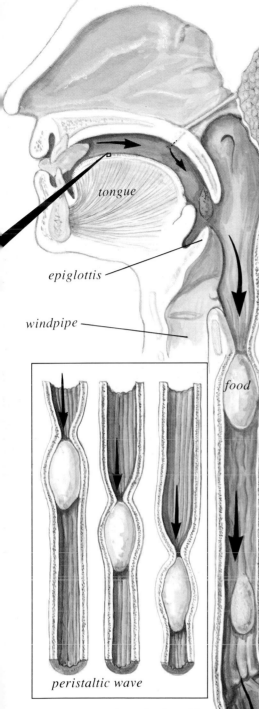

The pharynx

The pharynx forms a common channel for the digestive tract and the air passages behind the mouth and the nose. At the bottom it divides into the windpipe (*trachea*) and the gullet (*oesophagus*.). When we swallow, the epiglottis flattens to cover the windpipe, so that food is channelled into the gullet and not into the windpipe ('the wrong way'). Swallowing starts when the tongue and the muscles in the pharynx press the food towards the back and down into the oesophagus. This is a pipe about 25 cm long which conveys the food to the stomach. Gravity helps the food to go down. In addition, the muscular tissue in the gullet will contract at the precise point where the food is. This contraction goes from top to bottom and pushes the food through before it. This movement is called a *peristaltic wave*. The peristaltic wave is a reflex action which starts as we swallow and conveys the food into the stomach in about five seconds.

The stomach

The glands in the stomach (ventriculus) produce *gastric juice*. Production is controlled by hormones and nerve impulses. Gastric juice consists of water, enzymes and *hydrochloric acid*. Hydrochloric acid kills bacteria and is furthermore necessary to ensure that the enzymes can break down the proteins into smaller constituent parts. The inside of the stomach is covered with mucus so that the gastric juice does not break down the actual stomach wall. When there is too much gastric juice or too little mucus, a *stomach ulcer* may occur. This may occur in the stomach itself (*ulcus ventriculi*) or in the duodenum (*ulcus duodeni*). The stomach wall consists of smooth muscular tissue. Contractions in the muscular tissue mix the gastric juice with the food and knead the contents of the stomach. Then peristaltic waves gradually move the food into the duodenum.

tongue

epiglottis

windpipe

food

peristaltic wave

the pyloric sphincter muscle lets the stomach contents pass into the duodenum

stomach lining

mucous membrane/gastric mucosa (greatly enlarged)

stomach contents

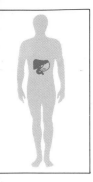

The digestive tract - middle section

The final break-down of the food occurs in the duodenum and in the rest of the small intestine. Bile, pancreatic juices and enzymes in the intestine wall are involved in converting the food into nutrients which are absorbed from the intestines.

Digestion in the duodenum

Fats cannot be dissolved in water and tend to coagulate in the intestines. When the volume of fat in the intestine is very large, *bile* is squeezed out of the gall bladder into the duodenum. The bile helps to divide up (emulsify) the fat into small globules. The globules can easily be further broken down in the intestine by enzymes. The bile consists of salts and pigments, *cholesterol* and water. Bile is produced in the liver and stored in the *gall bladder*. In the *gall bladder* some of the water is extracted so that the bile becomes more concentrated.

When the food is pressed out of the stomach into the duodenum, the nutrients have not yet been completely broken down. Therefore, the pancreas is stimulated to produce *pancreatic juice*. The pancreatic juice is excreted into the duodenum via the pancreatic passage. It contains sodium bicarbonate and enzymes and is weak basically in order to neutralize the gastric juice. The enzymes in the pancreatic juice and the digestive enzymes, which are produced by small glands in the intestinal wall, break down the nutrients so that they can be absorbed out of the intestines. The proteins are converted into amino acids, the carbohydrates into glucose and other tiny sugar molecules, and the fats are converted into fatty acids, glycerol and glyceride.

The tasks of the liver cells are:
- to convert proteins, carbohydrates and fats;
- to store glucose in the form of glycogen;
- to break down hormones, waste matter, the remains of blood cells, medicines and toxic materials;
- to produce bile.

For more information about the digestion, see:
The digestion (44, 45)
The digestive tract - upper section (46, 47)
The digestive tract - lower section (50, 51)

For more about insulin and glucagon, see:
Hormones (42, 43)

Apart from being an exocrine gland which secretes pancreatic juice into the intestines, the pancreas is also an endocrine gland which secretes the hormones insulin and glucagon into the blood.

The liver

The liver is the largest gland in the body (in adults it weighs about 1.5 kg) and has many important tasks. The liver cells remove matter from the blood and from the intestines. These are used as building bricks in order to make materials in blood proteins, which the body needs. Many waste products, toxic substances (*alcohol*, for example) and medicines are broken down by the liver cells so that these substances can be more easily removed from the body (for example, via the urine). Because the liver cells have so many tasks, it is important that the liver gets both oxygen-rich and nutrient-rich blood.

The blood supply

The blood supply travels to the digestive organs via branches of the aorta. From the arteries the blood passes into the capillaries in the mucous membrane. In the mucous membrane of the intestines the blood absorbs nutrients, salts, minerals, vitamins and water. Afterwards it is collected in the veins which open into the portal vein. This conveys the blood from the stomach and the intestines to the liver, supplying it with nutrient-rich blood. The liver obtains oxygen-rich blood from the hepatic artery, which is a branch of the aorta. In the liver the blood from the hepatic artery is mixed with the blood from the portal vein and then passes through a network of capillaries. In this way the liver cells obtain both oxygen-rich and nutrient-rich blood, which they need to perform their tasks. The portal vein differs from most other veins because it conveys the blood from one network of capillaries into another network. The body is provided with the correct amount of glucose (blood sugar) via the portal vein.

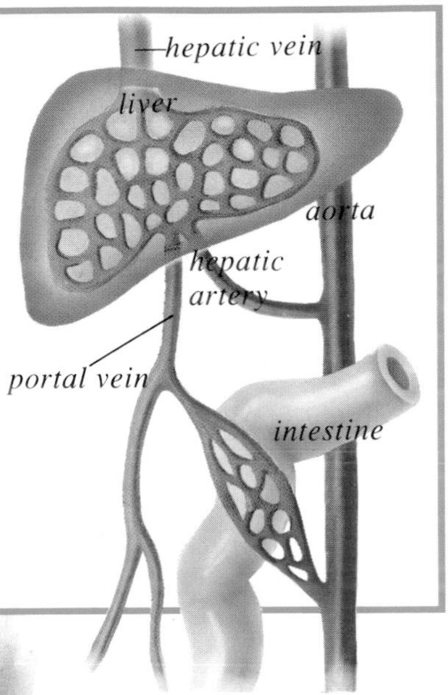

hepatic vein

liver

aorta

hepatic artery

portal vein

intestine

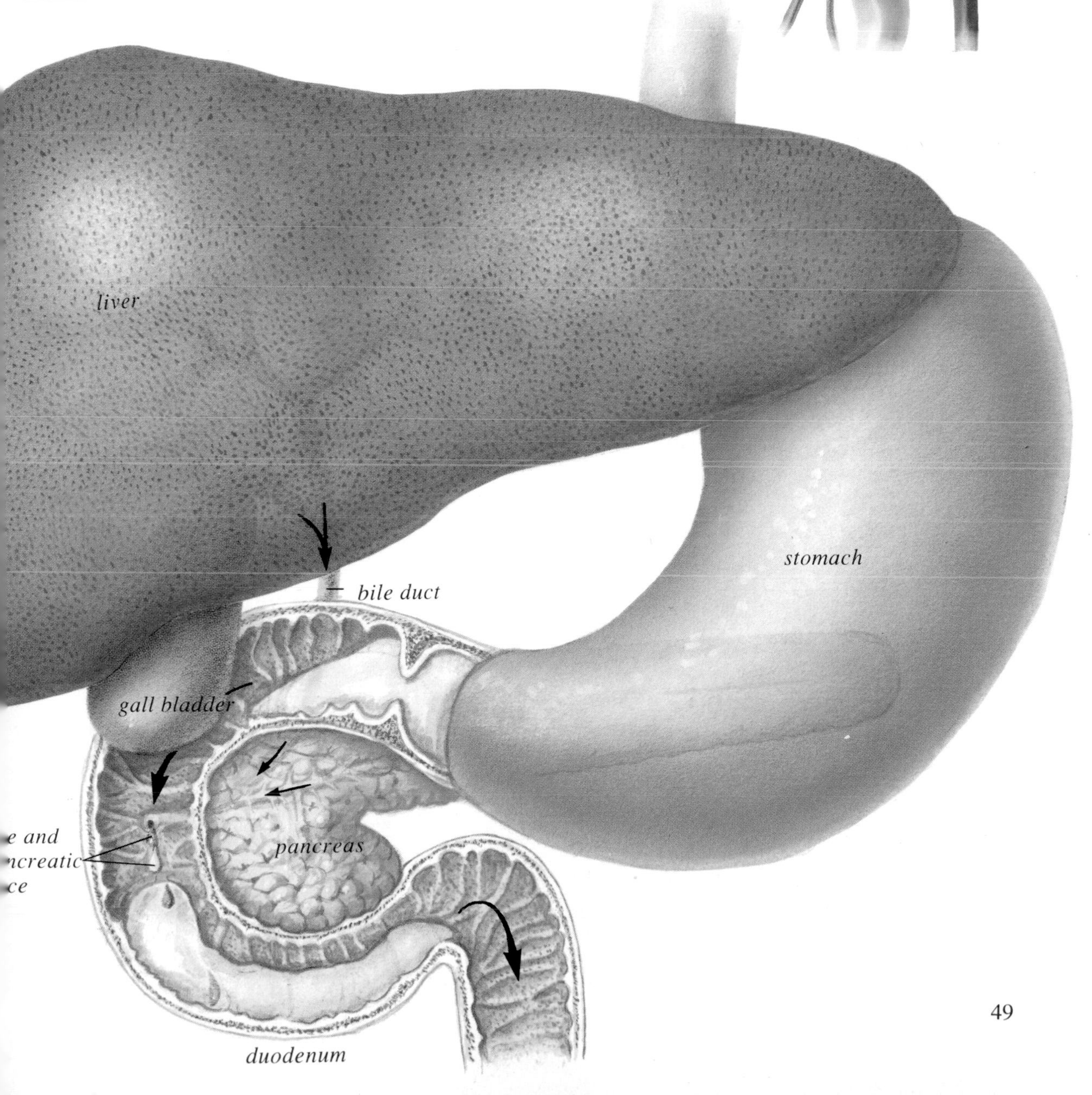

liver

stomach

bile duct

gall bladder

e and ncreatic ce

pancreas

duodenum

The digestive tract - lower section

Nutrients are absorbed from the small intestine. In the large intestine, further resorption of salts and water takes place, so that the faeces are condensed.

The small intestine

The small intestine is 3-4 metres long and 3-4 cm in diameter. The *duodenum* is the first part of the small intestine. The middle section is called the *jejunum*, the lower section the *ileum*.

In order to ensure that the absorption of nutrients occurs as effectively as possible, the inside of the small intestine has large folds. In addition the mucous membrane is covered with many small *intestinal villi*. On the surface of the epithelial cells there are many minute projections (*microvilli*). In this way the intestine has a very large active surface (totalling about 250 m^2!).
The nutrients (other than fats) are absorbed into the blood by capillaries in the villi. The epithelial cells in the mucous membrane transfer the nutrients from the intestine by a kind of pump (active transport) and this requires a great deal of energy. The fats are absorbed directly into lymph vessels, the largest of which, the thoracic duct, later transfers the fats to the bloodstream.

The intestinal wall consists of smooth muscular tissue. Muscular contractions knead the contents of the intestine and mix it with bile and pancreatic juice. The muscular contractions are transmitted gradually at regular intervals towards the end of the intestine in a peristaltic wave. In this way the contents of the intestine are pushed further along the intestine.

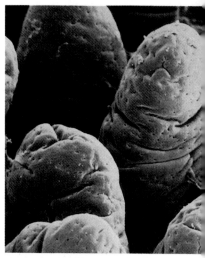

intestinal villi (greatly enlarged)

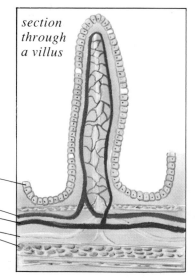

section through a villus

epithelial cells
vein
artery
lymph vessel
smooth muscular tissue

For more about the digestive system, see:
The digestive system (44, 45)
The digestive tract - upper section (46, 47)
The digestive tract - middle section (48, 49)

See also:
The lymphatic system (24)

The large intestine

The *large intestine (colon)* is about 1.5 m. long. It consists of an ascending section, a transverse section, a descending section and an S-shaped part. The large intestine reabsorbs water and controls the mineral balance. The absorption of water means that the loose contents of the intestine are rendered more concentrated. When the food passes through the large intestine too quickly, too little water is removed and the faeces are thin. When the food passes too slowly, a great deal of water is resorbed. Then the faeces become hard. The large intestine contains a great number of *bacteria*. Some of the bacteria fulfil a useful function, including vitamin production.

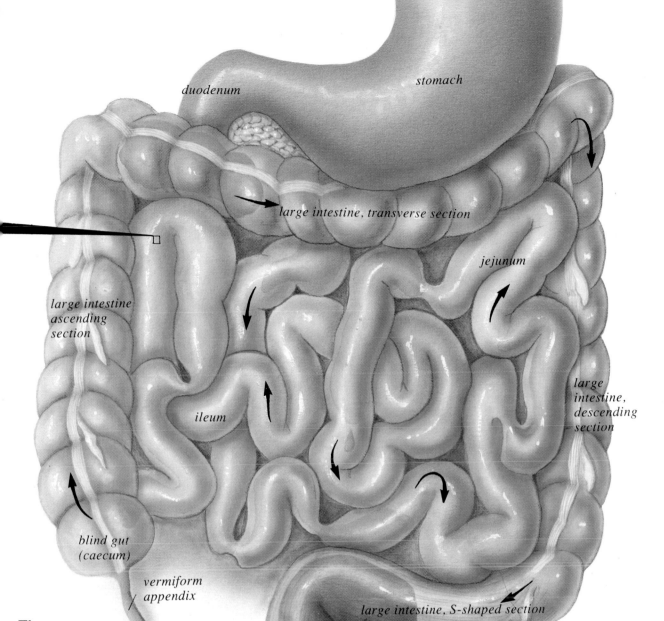

duodenum

stomach

large intestine, transverse section

jejunum

large intestine
ascending
section

large
intestine,
descending
section

ileum

blind gut
(caecum)

vermiform
appendix

large intestine, S-shaped section

rectum

The blind gut

The blind gut (*caecum*) is the beginning of the large intestine (*colon*).

The *vermiform appendix* hangs like a long, thin worm from the caecum. In humans the appendix has no particular task. *Inflammation of the caecum (appendicitis)* is in fact an inflammation of the vermiform appendix.

The rectum

Peristaltic waves reach the *rectum* once or twice a day. These try to push out the contents of the intestine. Half of the faeces is made up of water and unabsorbed food residue. The remainder consists of intestinal bacteria (both living and dead).

The rectum has two sphincter muscles. The inner one is a ring of smooth muscular tissue around the intestine. This is opened by the peristaltic waves. The outer sphincter muscle is a ring of transverse striated muscular tissue. This is voluntarily controlled, so that we are able to retain the faeces.

51

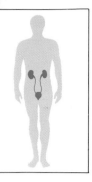

The urinary system

The urinary system consists of the kidneys, the ureter, the bladder and the urethra. It contributes to maintaining the water balance, the electrolyte balance and the acid/base balance. The urinary system excretes waste matter (such as urea, uric acid and other nitrogenous substances) via the urine. Apart from waste matter, urine contains water, salts, vitamins and hormones.

The water balance

More than half of our body consists of water. Many bodily functions require that this amount of water be kept constant. The amount of water in the body is determined by the amount of water provided by eating and drinking and the amount we discharge by urinating, sweating, defecating and exhaling. When there is little water in our body, the thirst centre in the brain is stimulated, so that we become *thirsty* and drink more. This adds to the amount of water. The amount of water can fall by increasing the discharge of urine. Urine production is controlled by the *antidiuretic hormone (ADH)*, which is secreted by the pituitary gland. ADH slows down the production of urine. When there is little water in the body, more ADH is made, so that the kidneys excrete less urine and keep the body at the right water level. When there is a lot of water in the body, ADH production drops, so that more water goes into the urine.

Alcohol slows down the production of ADH. *Alcohol consumption* increases the discharge of urine, so that the body becomes short of water. This again leads to thirst.

For more about electrolytes, see:
The blood (22, 23)

For more about the urethra, see:
The male sex organs (66, 67)
The female sex organs (68, 69)

See also:
Hormones (42, 43)

The electrolyte balance

Electrolytes are important for the body. They are provided by eating and drinking. The hormone *aldosterone* from the adrenal cortex controls the secretion of some electrolytes. The quantity of electrolytes and their inter-relationship is important for the normal functioning of the body's cells.

The acid/base balance

The body is dependent on a constant acid level. The kidneys contribute to the excretion of excess acids or bases.

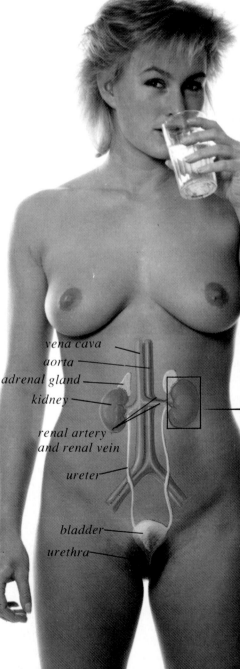

vena cava
aorta
adrenal gland
kidney
renal artery and renal vein
ureter
bladder
urethra

Urine production

Blood is carried via the renal artery to the arterioles in the two *kidneys*. In the kidneys there are systems of small ducts called *nephrons*. There are about 1 million nephrons in each kidney.

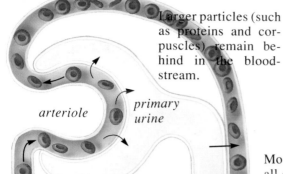

Larger particles (such as proteins and corpuscles) remain behind in the bloodstream.

arteriole

primary urine

Water and dissolved molecules are squeezed out of the arteriole into the nephrons. This fluid is called *primary urine*.

Each day about 180 litres of *primary urine* are produced. Only 1.5 litres of this is excreted as urine, the remainder is resorbed. The waste matter remains behind in the urine in concentrated form. The most important substances excreted are urea, uric acid and salts.

Most of the water and all of the nutrients in the primary urine go back into the blood via the nephrons (*resorption*).

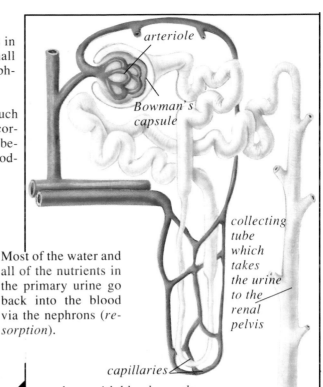

arteriole

Bowman's capsule

collecting tube which takes the urine to the renal pelvis

capillaries

a nephron with blood vessels

nal cortex

nal medulla

renal pelvis

ureter

The discharge of urine is dependent on blood pressure. If the blood pressure is too low, then urine production stops. The kidneys themselves are concerned with controlling blood pressure.

Urinating

The urine passes from the kidneys via the *renal pelvis* and the *ureter* to the *bladder* (*vesica urinaria*). This can usually hold about a third of a litre. As soon as the bladder is full enough, the smooth muscle tissue in the bladder wall contracts and tries to squeeze out the urine via the *urethra*. Then we feel the need to pass water and we can either urinate or retain it by the use of a voluntarily-controlled sphincter muscle. The *urethra* is shorter in women than in men.

Diseases

Bacteria which enter the urethra can cause infection of the urinary passages (*urinary infection*). Infection in the bladder, frequently caused by intestinal bacteria, causes *inflammation of the bladder* (*cystitis*), which is accompanied by frequent and painful urination and often by fever (pyrexia). Because women have a shorter urethra than men, they suffer more often from this disease. If the bacteria reach the renal pelvis *inflammation of the renal pelvis* (*pyelitis*) can occur.

In the case of diseases of, or damage to the urinary passages, other substances can usually be found in the urine (such as albumen, glucose, corpuscles and bacteria). Bacteria are visible with a microscope when analyzing the urine sediment.

The skin

The skin is the largest organ of the body. It covers the whole body and forms the boundary between the internal and external environment. The skin therefore has the task of protecting the body against external influences.

The skin protects

The skin prevents bacteria and other harmful organisms from penetrating the body. It protects against wear and tear, knocks and abrasions, and prevents penetration by chemicals and other substances. The skin ensures that the body does not get rid of too much water and other important materials. Small sensory organs in the skin register harmful influences and send information to the central nervous system.

See also:
The functions of the skin (56, 57)
Tissue (14,15)
Sensory nerve endings and nerve cells (30, 31)

The structure of the skin

The skin consists of three layers:
The outer skin, the *epidermis*, is made up of several layers of scaly epithelia. The epidermis prevents penetration into the body of microbes and chemicals. The cells in the epidermis are formed directly above the dermis and then gradually pushed upwards. During this upwards movement they die off and change slowly into a dense dead mass. This is called *keratin*.
The outermost keratin cells are shed. It takes about 24 days from formation for a cell to be shed. When the keratin cells are visible as white scales, we call it *scurf* or *dandruff*.

The *dermis* is made of connective tissue. This is extra strong and elastic and gives protection against mechanical influences. The dermis is richly provided with nerves and blood vessels.
The dermis gives young people a smooth, tight appearance. Over the course of years, the skin becomes less elastic and wrinkled.

The bottom (*subcutaneous*) layer of skin consists mainly of fat cells and forms an important layer of insulation, so that the body does not lose too much heat.

When the skin gets extra wear as it does in summer when we go barefoot, it thickens. This happens by the formation of more cells than are shed.
The palms of the hands and the soles of the feet are normally exposed to a lot of wear and tear. Therefore calluses form at those points.

Nails

The nails are made up of epithelial cells, which harden in almost the same way as the cells in the epidermis. The *keratin* is however much harder. The nail is formed in the *matrix* and then pushed outwards. As the keratin is dead material, it does not hurt when the nails are cut.

Hair

Hair is also formed by hardening of the epithelial cells, but in a different way from the skin and the nails. The hair is formed in follicles and then pushed outwards. Hair consists of dead cells. Therefore it does not hurt when the hair is cut.

In the *hair follicles* are pigment cells which supply the colour. The amount of pigment determines the colour of the hair. As we get older, the cells reduce the production of pigment and our hair turns grey. White hair has no pigment but does contain minute air bubbles.

Glands in the skin

The *sebaceous glands* produce *sebaceous matter* which is excreted via the follicles where the hair projects from the skin. The sebaceous matter (*sebum*) lubricates the skin and makes it soft and supple. The openings in the skin are called *pores*.

The *sweat glands* produce *sweat*, a fluid containing salts.

Skin flora

On the surface of the skin there is a flora of normally harmless bacteria, which serve, amongst other things, to combat more harmful bacteria.

Diseases

In *puberty* sebaceous production increases. In people with narrow pores, these can easily become clogged. This can be accompanied by inflammation of the sebaceous glands, causing *pimples* (*acne*).

The functions of the skin

Apart from protecting the body against external influences, the skin is also an important aid to keeping the body informed about the environment. In addition, the skin plays a central role in regulating temperature, so that the body temperature is maintained at around 37°C.

Signals from the skin

The skin and the fine muscles under the skin are able to transmit signals to other people. Most of these signals we receive unconsciously. Think what feelings are evoked by the soft creamy skin of a baby or by the skin of someone to whom we feel attracted. Skin contact gives most people a positive feeling.

Red bloom on our cheeks and an even brown tint indicate that we are healthy, whilst a particularly pale skin may be a sign of illness.

It is not only the appearance of the skin that provides information. Sweat glands secrete substances which we unconsciously smell and interpret. The secretion is above all particularly strong in the case of sexual attraction. This may be one of the reasons why we 'feel' that someone does or does not like us.

For the mucous
membrane, see:
**The digestive system
(44, 45)**

See also:
**The skin (54, 55)
Sensory nerve endings
and nerve cells (30, 31)
Heart and blood vessels
(18, 19)**

Sunlight stimulates the skin to produce *Vitamin D*. In addition sensible sunbathing leads to us feeling healthier and fitter. Excessive sunbathing is, however, harmful to the skin, particularly when the skin has little protective pigment.

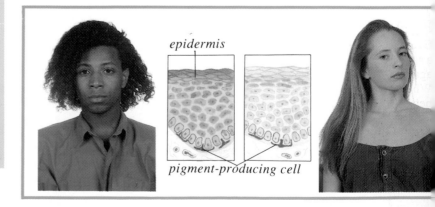

epidermis

pigment-producing cell

Body temperature

The skin plays an important role in controlling body temperature.

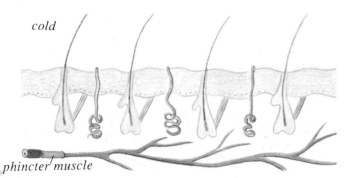

cold

sphincter muscle

When it is *cold*, we want to keep our body warm. Small circular sphincter muscles in the blood vessels in the skin contract and limit the blood supply, so that we do not lose too much heat from the blood via the skin. The fat in the subcutaneous layer acts as insulation. Then the tiny muscles on the hair follicles of the downy hair tighten, so that the hairs become erect (*goose pimples*). In this way the body endeavours to maintain a stable, insulating layer of air.

heat

When it is *hot*, we want our body to cool down. The sphincter muscles in the blood vessels in the skin relax, and the blood supply increases, so that a lot of heat can escape through the skin. The downy hairs lie flat so there is no longer an insulating layer of air. The sweat glands produce sweat. As it evaporates, the skin cools down.

Diseases

The skin provides information to the outside world and it is clear that negative signals are emitted by people with a visible, but harmless skin disease. Therefore, it is important to take into account how people live with this disease and not just how serious the disease is regarded medically.

Pigment

Everywhere underneath the epidermis there are cells which produce a special brown colouring (*pigment*). This pigment protects the body against harmful rays, particularly against ultraviolet rays in sunlight. When we sunbathe, pigment production increases, so that the skin is better able to withstand the sun's rays. The pigment gives the skin a browner tinge.

Some people over the course of time have been more exposed to *ultraviolet radiation* than others. In these people the skin has adapted to the circumstances in temperate climates, by a greater production of pigment than is the general rule, so that they have darker skins.

Healing of wounds

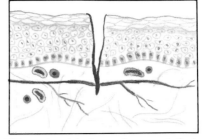

When the skin is so damaged that the cells die, wounds occur. They are occasioned by mechanical injuries, extreme heat or cold, strong irradiation or poor blood circulation. Usually blood vessels are also damaged, so blood leaks into the wound.

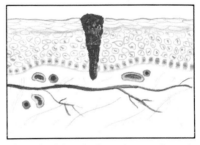

First the blood clots (*coagulates*), so that the bleeding stops. Then white blood corpuscles consume the clotted blood and the dead cells.

In the area of the wound, connective tissue cells divide more quickly than normal in order to replace the dead cells.

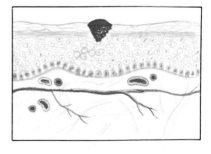

Then the epithelial cells grow over the wound and so a new dermis and epidermis are formed. If the wound was a large one, the new skin will look different from the old skin. We then speak of a *scar*.

57

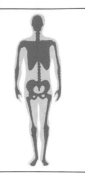

The skeleton

The skeleton provides shape for the body. It supports and protects the other organ systems and is part of the motor apparatus. The human skeleton consists of around 200 bones.

The different types of bones

The bones are constructed of two kinds of bone tissue, the *compact bone tissue* (*substantia compacta*) and the *spongy tissue* (*substantia spongiosa*). The bones are covered with a connective tissue membrane, the *periosteum*, and with nerves and blood vessels. The latter provide for the supply of nutrients.

There are two types of bones: long bones and flat bones. *Long bones* are long and slender with a long shaft (*diaphysis*) and two *extremities* (*epiphyses*). The extremities consist of spongy bone tissue with compact bone tissue on the outside. Where the extremities serve as joint surfaces, they are covered with *cartilage*. The shaft is hollow and is made of compact bone tissue. The hollow cavity (*medullary canal*) contains bone marrow. Between the epiphysis and the shaft there is a *growth plate*. Long bone extremities are light but strong. The bones in the upper arm and the thigh are examples of these.
Flat bones consist of two thin plates of bone of compact tissue with spongy tissue between them. These are found in the skull, among other places, but the ribs and the breast bone too are flat bones. The flat bones in the wrists and ankles and the vertebrae are also called *irregular bones*.

For more about height growth, see:
Puberty (64)

See also:
The joints (60)
Tissue (14, 15)

For the names of the largest bones, see:
Anatomical terms (86)

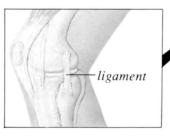

ligament

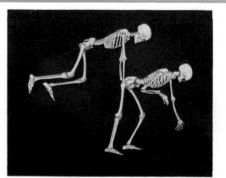

The skeleton is held together by strong connective tissue called *ligaments*. Without the ligaments the skeleton would collapse. The shape of the joint surface together with the ligaments determines what movements are possible for the joint.

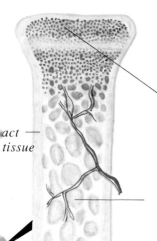

act tissue

In the medullary canal and in the spongy bone tissue we find bone marrow. There are two types of bone marrow:
Red bone marrow consists of blood-producing tissue. We find red bone marrow in the spongy bone tissue.

Yellow bone marrow consists mainly of *fat*. Yellow bone marrow can be found in the medullary cavity.

growth plates (before puberty)

'sealed' growth plates (after puberty)

Nutrient requirements

Bones consist of living tissue, which just like all other tissue needs oxygen and nutrition. They get this via the blood vessels which are found in the bone tissue.

There is a lot of calcium in bones. Calcium is also important to other parts of the body. If too little calcium is consumed in food, calcium is withdrawn from the bones and carried in the blood. The bones therefore become brittle. Vitamin D plays an important part in the exchange of calcium between the blood and the bones. When there is a lack of Vitamin D in the body, not enough calcium is absorbed from the intestines and supplied to the bones, so that they become more liable to break.

Older people's bones rapidly become brittle. This is caused by too little body movement, especially if they are confined to bed.

When the body grows

Bones grow in width, because bone tissue is deposited on the outside of the bone, whilst simultaneously tissue on the inside of the bone is consumed.

Bones grow in length, because on one side of the growth plates, which consist of cartilage, new cartilage is formed and on the other side the cartilage is transformed into the bone tissue of the shaft.

Growth of the bone length stops at the end of the teen years, so that on x-rays, growth plates can no longer be seen. They are totally solidified into bone tissue.

Injuries

Bone fracture is a very common injury. There are many types of fractures. When the bone is out of position, the doctor has to put it back in its proper place. The break heals by new bone tissue being deposited around the fracture, so that the bone thickens at that point. When the bone has grown together again completely, other cells will consume the excess bone tissue.

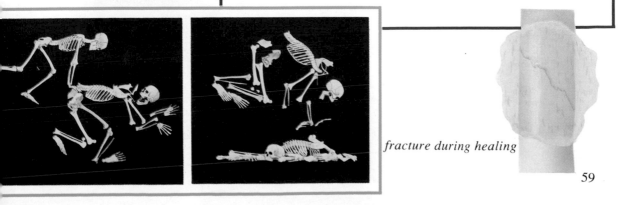

fracture during healing

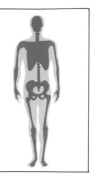

The joints

In order to hold the bones securely together, there are joints between them. Most of these joints are constructed in such a way that the bones can move in relation to each other.

We differentiate between three main types of joints: true articulated joints, fibrous joints and cartilaginous joints.

Articulated joints

When we move an articulated joint, we move the bones in relation to each other. Only with true articulated joints can we make clearly visible movements.

There are different types of articulated joints. The names are taken from the appearance of the end of the bones (ball-and-socket joint, saddle joint) or from the movement made by the bones in relation to each other (hinged joints).

In articulated joints, the ends of the bones are covered with cartilage. Between this is the *synovial membrane (synovium)*. The cartilage and the synovial membrane ensure that there is little friction between the bones as the joint is moved. Around the joint is a thick *joint capsule* which produces the synovium and prevents the synovium from escaping. On the outside are the *lateral ligaments*, which hold the bones together. Both the shape of the bones and the lateral ligaments determine the limits within which the articulated joint can move.

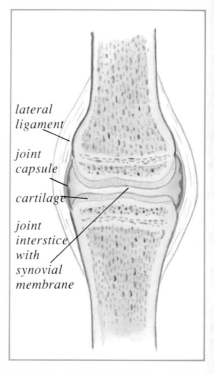

lateral ligament

joint capsule

cartilage

joint interstice with synovial membrane

For more about nerves, see:
The peripheral nervous system (36, 37)

See also:
The skeleton (58, 59)
Tissue (14, 15)

For the names of the individual important articulated joints, see:
Anatomical terms (86)

The hip joint is a ball-and-socket joint.

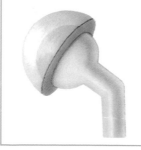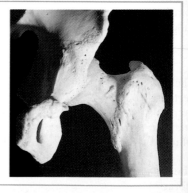

Toe joints are hinged joints.

Fibrous joints

The joints between the bones of the skull are examples of *fibrous joints*. Here an irregular serrated edge holds the bones together. Such a joint offers no possibility of movement.

Cartilaginous joints

The uppermost joint in the breastbone is an example of a *cartilaginous joint*. Here the cartilage and the ligament hold the bones together. This joint offers little possibility of movement.

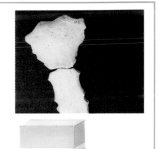

he knee joint is ie largest and ost complicated oint in the body. here are two ar- culatory disks of *artilage* in the nee joint. These re called the *enisci*. Besides ie usual lateral gaments around ie knee joint we lso find in the nee articulation vo small articular gaments called ie *cruciate liga- ents*. oth the menisci nd the cruciate gaments con- ibute to the knee oint's normal unctioning.

spinal cord

spinal nerve

interstitial disk

he *spinal column* consists of 26 pieces of bone (7 cervical ertebrae, 12 dorsal vertebrae, 5 lumbar vertebrae, the acrum and the coccyx). Between each vertebra is an iterstitial disk* of cartilage and some small saddle joints hich allow the vertebrae to move in relation to each other.

he muscles and articular ligaments attached to the ver- brae limit the forwards, backwards and sideways iovements which may be made by the spine. We are lso able to rotate the spine.

here is a central hole in each vertebra. Together these oles form the *vertebral canal*. The spinal cord runs rough the vertebral canal and the *spinal nerves* pass out etween the vertebrae.

Injuries

If the spinal column is sub- jected to an abnormal load, the wall of a disk may crack, and the soft core may be displaced and press on a nerve. This is what happens with a *slipped disk (prolapse of an interver- tebral disk – PID)*.

The muscular system

Muscles are able to contract, which enables the body to move. We have three kinds of muscular tissue, striated muscular tissue, cardiac muscle tissue and smooth muscular tissue. The striated muscular tissue allows voluntary movements to be made.

Striated muscular tissue

The skeletal muscles are *voluntary* muscles which are attached to the skeleton. We have more than 600 of these muscles which vary in size. Each muscle can move a certain part of the body. Some muscles are very strong, others have more of a capacity for stamina. Both strength and stamina can be increased by training.

The muscles obtain energy in order to work (to contract) by burning up nutrients. Oxygen is necessary for this. During heavy, prolonged work the muscles are unable to completely metabolize the nutrients because they receive too little oxygen. Then the waste product *lactic acid* is formed. When lactic acid accumulates in the tissue, the muscles become painful.

For more about the different types of muscular tissue, see:
Tissue (14, 15)

See also:
Coordination between the organ systems during physical exertion (84, 85)

For the names of the most important muscles, see:
Anatomical terms (87)

Muscle origin

The muscles are fixed to the skeleton by a *tendon* of connective tissue.

In many places *bursas* are to be found, which protect the tendons against wear and tear as they slide over the projections on the bones.

Muscle insertion

At the muscle insertion point the muscles are fixed to the skeleton by a tendon. In order to move any part of the body, the muscles must have their origin on one side of a joint and their insertion on the other side.

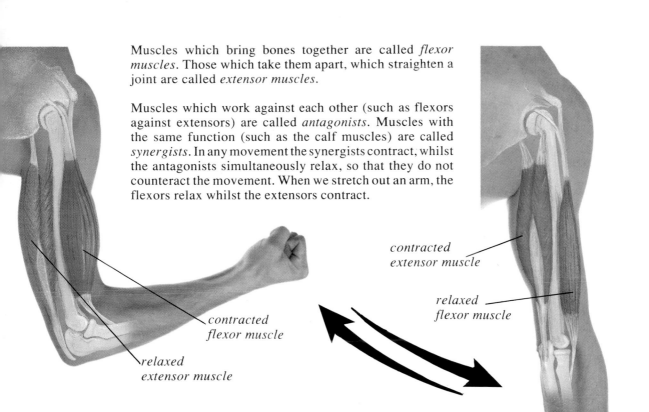

Muscles which bring bones together are called *flexor muscles*. Those which take them apart, which straighten a joint are called *extensor muscles*.

Muscles which work against each other (such as flexors against extensors) are called *antagonists*. Muscles with the same function (such as the calf muscles) are called *synergists*. In any movement the synergists contract, whilst the antagonists simultaneously relax, so that they do not counteract the movement. When we stretch out an arm, the flexors relax whilst the extensors contract.

contracted extensor muscle

relaxed flexor muscle

contracted flexor muscle

relaxed extensor muscle

The movement of any part of the body consists of movements of several joints. A great many muscles may be involved. When we kick a ball, all the muscles which straighten the knee and the ankle contract, whilst the knee flexors and the ankle flexors relax.

The cardiac muscles and smooth muscular tissue

The *cardiac muscles* contract to pump the blood round the body. The *smooth muscular tissue* can be found in many places, including the walls of the digestive tract, the bladder, the womb and the blood vessels. Annular constrictors or ring-shaped sphincters of smooth muscular tissue can open and close the blood vessels, so that the blood is distributed to different parts of the body.
Smooth muscular tissue transports food, kneads it in the intestines, squeezes out the urine as we urinate and expels the baby during birth.

Muscular cramp and tension

Overloading of the muscles frequently occurs after great exertion. This can cause painful inflammation in and around the tendons. Inflammation of the tendons is also not unusual. *Muscular cramp* is a prolonged, painful contraction of the muscles. This may have many causes and in most cases is not connected with disease.

Muscular tension often occurs under stress. The tension frequently happens unconsciously. It can be very painful and is one of the most common complaints of the muscular system.

Puberty

During puberty we develop from a child to an adult. This brings with it a number of social changes. However, the most visible change is that of physical development; from a child to an adult, sexually mature body.

Reaching sexual maturity

The specific characteristics of the two sexes are called *sexual features*. Amongst these the sex organs are the *primary sexual features*. We call the other differences the *secondary sexual features*.

The changes during puberty start in the early teen years in girls, often somewhat earlier than in boys. These changes, which occur both in the sex organs and in the rest of the body, are mainly caused by the influence of hormones. These are in turn controlled by the brain. The pituitary gland starts producing two types of hormones, *FSH* (follicle stimulating hormone) and *LH* (luteinizing hormone). In boys, these hormones trigger the production of sperm cells and testosterone. In girls they stimulate the ripening of the ova and the production of oestrogen in the ovaries.

Menstruation is dealt with as a separate subject (70, 71)

For more information about the sex organs, see:
The male sex organs (66, 67)
The female sex organs (68, 69)

For more about speech, see:
Speech and breathing (28, 29)

See also:
Chromosomes and cell division (12, 13)
Hormones (42, 43)

Sex hormones

Oestrogen (from the ovaries) and *testosterone* (from the testicles) are called sex hormones. These hormones are also produced, in small quantities, in the adrenal glands, so that everybody possesses both hormones. During puberty the primary and secondary sexual features develop through the influence of the sex hormones. They are also responsible for sexual instincts.

Testosterone stimulates the growth of hair around the genitals and under the arms in both girls and boys. In boys, hair grows on the face and the larynx enlarges so that the sound of the voice changes (*the voice breaks*).

Oestrogen in girls makes the breasts grow and develop. Boys also have the structure for breasts but they do not develop because boys do not have enough oestrogen in their bodies.

The sex hormones influence muscle growth and fat distribution in the body, which gives girls a feminine and boys a masculine appearance.

Menstruation

Menstruation in girls begins in puberty. Menstruation is closely connected with ovulation, but usually begins before ovulation really gets started. To begin with, therefore, menstruation is quite irregular.

The body grows

Growth in height during puberty is particularly rapid and then stops almost completely. Because the growth in height (and with it, puberty) stops earlier in girls, they are not on average as tall as boys.

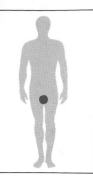

The male sex organs

The main task of the male sex organs is the production of sex cells and the transfer of these cells to the female sex organs, where fertilization can take place.

The penis

The penis (the male member) is constructed from three *corpora cavernosa* (or *erectile tissue* structures) and covered with skin. The corpora cavernosa are fastened at the base of the pelvis and run through the *shaft* of the penis to the glans. The *glans* is covered with a particularly thin skin and is highly sensitive. An extra fold of skin, the *foreskin* (*prepuce*), often lies over the largest part of the glans and protects it.

The corpora cavernosa consists of hollow spaces which can be filled with blood. During sexual arousal a great deal of blood flows into this erectile tissue. Then the penis becomes stiff and erect, so that it can be inserted in the vagina during intercourse.

The scrotum

The *scrotum* protects the *testicles*. In order to function, the testicles must be at a temperature of slightly under 37°C. When it is cold, the smooth muscular tissue in the scrotum retracts the testicles against the body in order to keep them warm. When it is warm, the smooth muscular tissue relaxes and the testicles are suspended free of the body.

The seminal vesicles

The two *seminal vesicles* produce a fluid in which the sperm cells can actively move about.

The prostate

The *prostate gland* produces a fluid which complements the function of the seminal vesicles. The fluid from the prostate and the seminal vesicles combined with sperm cells is called semen. Mucous glands lubricate the urethra and neutralize any urine residue.

The seminal duct

A seminal duct (*vas deferens*) transports the sperm cells from the epididymis to the urethra.

The epididymis

The *epididymis* consists of a collection of ducts in contact with the seminiferous tubules in a testicle. The sperm cells which are formed in the testicle are conveyed to the two epididymi via the seminiferous tubules. The sperm is stored there for a time.

See also:
Reproduction (72, 73)
Puberty (64)

About AIDS, see:
The lymphatic system (25)

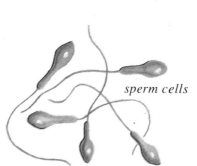

sperm cells

The testicles

The testicles have two main tasks: the production of billions of *sperm cells* (*spermatozoa*) and the production of the male sex hormone *testosterone*. In each of the two testicles there is a network of small ducts, the *seminiferous tubules*.

In the embryonic stage, the testicles lie up against the rear abdominal wall. They slowly move downwards through the abdominal cavity via the *inguinal canal*, through the abdomen wall and then descend into the scrotum. Sometimes at birth they are not fully descended. This can be assisted by surgical intervention or with the aid of medicines (hormones).

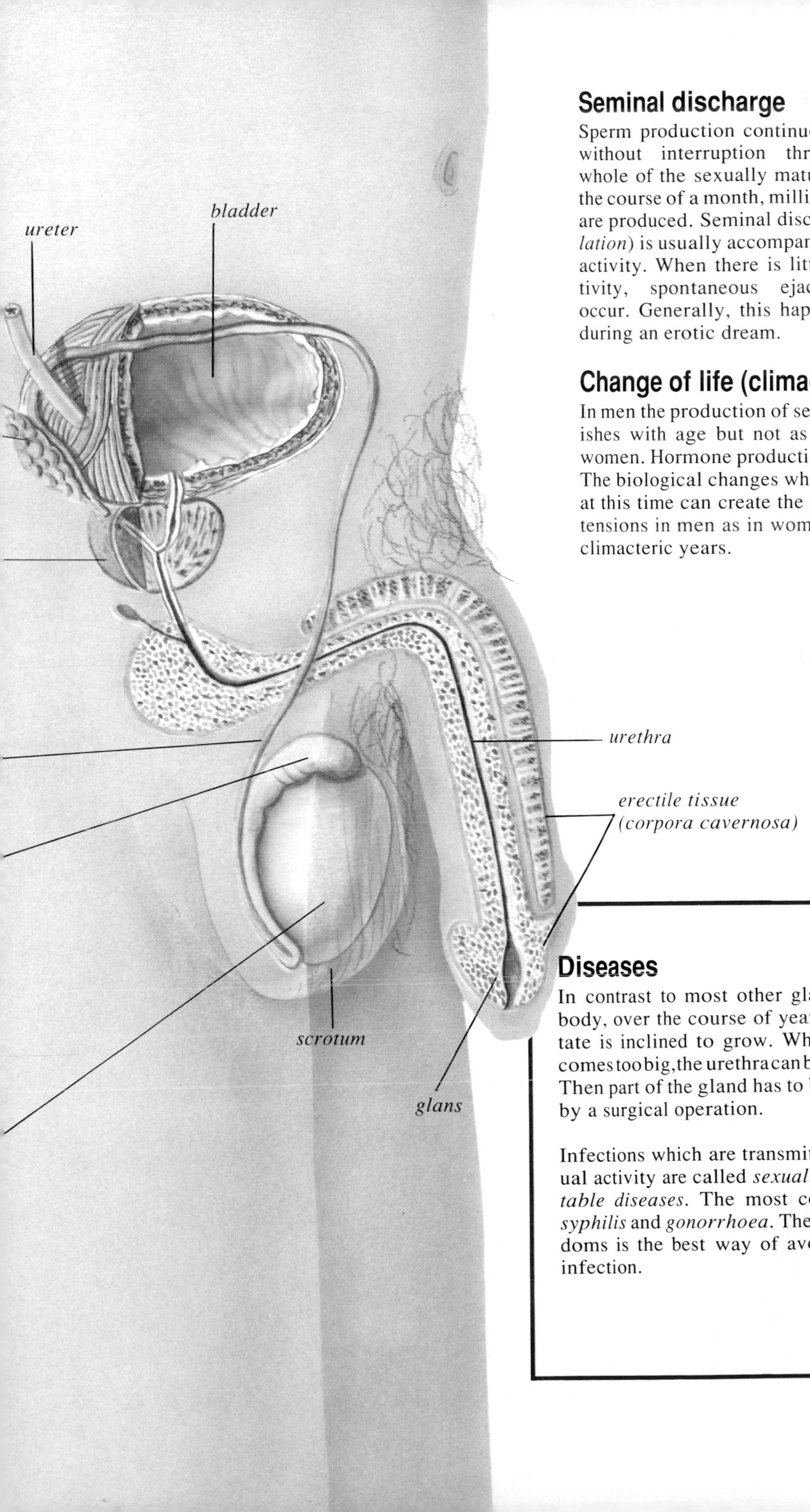

ureter

bladder

Seminal discharge

Sperm production continues practically without interruption throughout the whole of the sexually mature period. In the course of a month, millions of sperms are produced. Seminal discharge (*ejaculation*) is usually accompanied by sexual activity. When there is little sexual activity, spontaneous ejaculation can occur. Generally, this happens at night during an erotic dream.

Change of life (climacterium)

In men the production of sex cells diminishes with age but not as clearly as in women. Hormone production is reduced. The biological changes which take place at this time can create the same nervous tensions in men as in women during the climacteric years.

urethra

*erectile tissue
(corpora cavernosa)*

scrotum

glans

Diseases

In contrast to most other glands in the body, over the course of years the prostate is inclined to grow. When this becomes too big, the urethra can be occluded. Then part of the gland has to be removed by a surgical operation.

Infections which are transmitted by sexual activity are called *sexually transmittable diseases*. The most common are *syphilis* and *gonorrhoea*. The use of condoms is the best way of avoiding such infection.

The female sex organs

The female sex organs are formed in such a way that an ovum can be fertilized and will then develop in the womb (or uterus) into a new human being.

The breasts

The *breasts (mammae)* have the task of producing nutritious milk for the newborn baby. The breasts consist mainly of fat. The mammary glands are already present but enlarge and develop during pregnancy. Directly after the birth they begin to produce milk. The nipples have many sensory cells and these play an important role in suckling.

The vagina

The wall of the *vagina* consists of smooth muscular tissue and many sensory cells. The inside is coated with a strong mucous membrane. There are no glands in the mucous membrane but it is kept moist by mucus from the neck of the womb and by moisture which percolates into the mucous membrane from the blood vessels.
The vagina contains numerous harmless bacteria. These form lactic acid which keeps harmful bacteria and infections out of the area.

The clitoris

The *clitoris* is a small, conical organ containing many sensory cells. The clitoris passes under the skin into two corpora cavernosa, similar to the erectile tissue in the penis. During sexual activity, this erectile tissue is filled with blood so that the clitoris stiffens and becomes extra sensitive for sexual stimulation. Stimulation of the clitoris can lead to sexual arousal and orgasm.

The labia

The *labia majora* are covered with the usual skin plus hair, whilst the *labia minora* are covered by a thin, smooth layer of skin. The labia contain many sensory cells. Underneath these are glands. The sensory cells and the glands play a large part in sexual activity.

See also:
Puberty (64)
Menstruation (70, 71)
Reproduction (72, 73)

Sexually transmitted diseases are dealt with under:
The male sex organs (67)

The fallopian tubes

The two *fallopian tubes* are wide and in-fundibular (funnel-shaped) at the top in order to receive the egg cells (ova). They convey the ova from the ovary to the womb. The inside of the fallopian tubes is lined with a mucous membrane with moving outgrowths (or *cilia*,) which assist in transporting the ova.

The ovaries

The *ovaries* have two main tasks: the production of *egg cells* (*ova*) and the production of female *sex hormones*. Each of the two ovaries contains the rudiments of millions of ova, which can develop into mature egg cells. Once a month, in a sexually mature women, an ovum leaves each ovary.

The womb

The *womb* (*uterus*) is the size and shape of a pear. In the womb the unborn fetus is nurtured and fed until it develops into a viable human being.

The womb consists mainly of smooth muscular tissue. The inside is lined with mucous membrane rich in blood vessels and glands. Glands in the *neck of the womb* produce mucus which prevents infection from spreading from the vagina into the uterine cavity. The lowest opening of the womb is called the *cervix*.

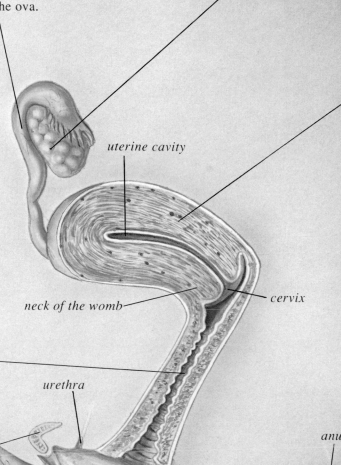

uterine cavity

neck of the womb

cervix

urethra

anus

Diseases

One very common and very irritating disease is a vaginal infection. This causes itching, burning pain and discharge.

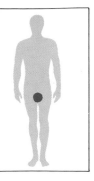

Menstruation

In women, a mature egg cell is released once a month. The mucous membrane of the uterus prepares for the implantation of the fertilized ovum. If there has been no fertilization, the membrane is rejected during menstruation.

Duration of menstruation

A menstrual cycle starts on the first day of menstruation and lasts until the first day of the next one. Ovulation takes place about 14 days before menstruation starts. On average the menstrual cycle takes 28 days, but may differ from one woman to another. Particularly at the start of puberty and during the menopause, considerable variations in timescale may occur.

Ripening of the ovum

In the embryonic phase, a start is made with the formation of millions of egg cells. In the two ovaries each month a few of the ova develop further, but usually only one ovum becomes fully mature.

See also:
The female sex organs (68, 69)
Reproduction (72, 73)
Embryo and fetus (74, 75)
Puberty (64)

Ovulation

When the ovum is mature, the *follicle,* the fluid-filled space around the egg, cracks open and releases the ovum.
When the egg is released (ovulation), the body temperature rises slightly. Some women experience a sharp pain on ovulation.

Ovulation is controlled by the *pituitary hormones FSH* and *LH.*

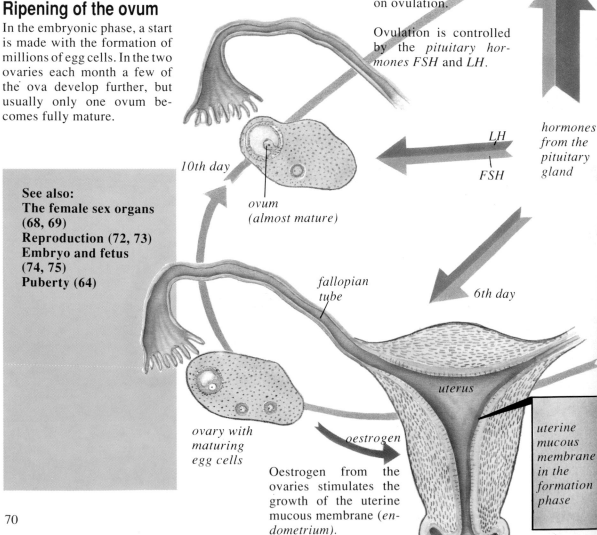

14th d

10th day

ovum (almost mature)

LH

FSH

hormones from the pituitary gland

fallopian tube

6th day

ovary with maturing egg cells

oestrogen

uterus

Oestrogen from the ovaries stimulates the growth of the uterine mucous membrane (*endometrium*).

uterine mucous membrane in the formation phase

ovum passing along the fallopian tube

thick uterine mucous membrane (endometrium) ready to receive the egg

yellow body

oestrogen

20th day

progesterone

vein

artery

gland

The yellow body

After ovulation the cells remaining in the follicles grow inwards and form the *yellow body* (*corpus luteum*).

The corpus luteum produces the hormone progesterone. Production of oestrogen in the developing follicles is continuous.

The supply of hormone by the corpus luteum enables the uterine endometrium to receive, protect and subsequently nourish the fetus.

Menstruation

When no fertilization has occurred, the corpus luteum ceases hormone production after 10-12 days. As a result, the blood supply to the major parts of the endometrium comes to a stop and the cells in those parts of the mucous membrane die off. After that the blood supply starts up again and large parts of the mucous membrane are rejected. Then *menstruation* begins. The blood and the remainder of the dead membrane flow out via the vagina as menstrual discharge. Enzymes ensure that the blood does not coagulate. Bleeding usually stops after 3-7 days.

28th day

1st day

ovum (in formation)

1st day

the endometrium becomes thinner during menstruation

The change of life

From about the 45th year, oestrogen production by the ovaries slows down. Menstrual bleeding becomes irregular. Then it stops completely. This time of life in a woman is called the *change of life* (*climacterium*). The change comes to an end after a few years at about the age of 50. These changes and the cessation of the reproductive ability may be accompanied by hot flushes, heavy sweating and sudden nervousness. These phenomena disappear after a year or so. The main aim of the change of life is to protect the body, so that it is not subjected to pregnancy when it is physically starting to get too old to cope with it. The phase of life after the final menstruation is called the *menopause*.

Many women frequently feel unwell before menstruation. This is due to the influence of the sex hormones on the body in general and on the mental and physical well-being in particular.

The actual menstrual bleeding can often be very painful. This is due, amongst other reasons, to small amounts of hormones released into the uterus causing contraction of the uterine muscles.

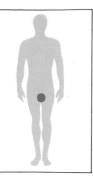

Reproduction

When a sperm cell fuses with an ovum after sexual intercourse, a new life is created. This individual will get half of its 46 chromosomes from each of its parents.

Sexual intercourse

The original aim of sexual intercourse is the creation of new life. It is also instinctively the most intense and intimate form of contact between two people. So that people can enjoy these feelings without intercourse leading to pregnancy, there are various means of contraception, such as the pill and the condom.

Sexual intercourse is introduced by *foreplay*. Visual impressions, odours and skin contact direct impulses to a nerve centre in the spinal cord, the *erection centre*. This centre has the task of preparing the body for copulation.

In men, blood is pumped into the corpora cavernosa so that the penis swells and becomes stiff and erect (*erection*). In women blood is pumped into the corpora cavernosa in the clitoris, so that this swells and stiffens. Moisture is provided by the membrane in the vagina, so that the surface is smooth. Furthermore, accumulations of blood make other places extra sensitive, such as the labia of the vulva, breasts, nape of the neck and the base of the spine. The stiffened penis is then introduced into the vagina. As the penis moves against the wall of the vagina and the clitoris,

highly sensitive sensory cells in these areas are stimulated. Then a great number of impulses are sent to a nerve centre in the spinal cord, the *orgasmic centre*. When it has been sufficiently stimulated, actual *orgasm* occurs.

In men this leads to a rhythmical contraction of the muscular tissue in the pelvic floor, particularly around the urethra. Sperm cells pass from the testicles via the epididymis and the genital ducts to the urethra, where they mix with the fluid from the prostate and the seminal vesicles. Then the sperm is pumped out through the urethra. This is called *seminal discharge (ejaculation)*.

In women too, orgasm creates a rhythmic contraction of the muscular tissue in the pelvic floor, particularly in the walls of the vagina. Orgasm can give a very great sense of pleasure to both men and women.

See also:
**The female sex organs
(68, 69)
The male sex organs
(66, 67)
Chromosomes and cell
division (12, 13)
Puberty (64)
Heredity and environment
(78, 79)**

Orgasm is over in a few seconds. In men it is usually some time before a further erection can be achieved, whilst in women a fresh orgasm is possible very rapidly. Experience of orgasm can vary from time to time and from person to person.

Women often need a longer and more powerful stimulation than men in order to achieve orgasm. Many women never achieve orgasm during intercourse but only by extra stimulation of the clitoris.

A number of physical and mental factors can lead to the biological reaction during foreplay and intercourse being weakened or failing to occur. This may be for any reason, from illness to nervousness, psychological pressure or a busy, hectic working day. Reduced sexual drive in men is called *impotence*. In women, it is termed *frigidity*.

With *masturbation* (self-gratification or onanism), an orgasm can occur without there being any intercourse.

Reduction cell division

A new individual is created when the sperm cell fuses with the ovum. Each cell in the new individual must have *46 chromosomes* (23 pairs). One chromosome of a pair comes from the father's sperm, the other from the mother's ovum. Therefore the sex cells always have 23 chromosomes, half the number of a normal cell. When the sex cells are created, they only need one chromosome of each pair. This is called *reduction cell division* (*meiosis*).

1 Cell before cell division. (To simplify the drawing, only 2 of the 46 chromosomes are illustrated, one pair of chromosomes of the 23 pairs.)

2 Each chromosome makes a copy of itself to make a 'double chromosome'. During this phase the double chromosomes of one pair lie close together and can exchange hereditary material (DNA).

3 The cell divides. From each pair, one 'double chromosome' moves to its own cell.

4 Two cells are formed, each with half as many 'double chromosomes' as the original cell.

5 Each new cell divides again. The 'double chromosomes' split up and one chromosome goes to each cell.

6 From the original cell, four cells have now been formed with half a set of chromosomes (23 chromosomes of which one is a sex chromosome).

The male sex cells develop into spermatozoa. Half the number of sperm cells contain an X-chromosome, the other half contain a Y-chromosome.

The female sex cells are called ova. Of the four cells which have been created by reduction division, only one will become a mature ovum. An ovum always contains one X-chromosome as its sex chromosome.

Embryo and fetus

From the time of fertilization, it takes about nine months until the new person is born. During this period the fetus obtains nourishment and oxygen from the mother's blood via the placenta. The fetus is connected to the placenta by the umbilical cord.

Fertilization

When sexual intercourse takes place around the time of ovulation, fertilization may follow. Intercourse can hasten the time of ovulation so that there is a greater chance of fertilization.

The sperm cells swim up through the uterus to the fallopian tube, where they meet the ovum. They combine in order to achieve fertilization, during which only one sperm cell will merge with the ovum to create a new individual. A fertilized ovum is called a zygote.

1st Week
Period of cell division

Directly after fertilization the zygote begins to divide. During the first week the cells divide many times and a cluster of cells is formed which moves down through the fallopian tube and then lodges itself in the endometrium.

cluster of cells

fallopian tube

cluster of cells lodges in the endometrium

fertilization

ovary

See also:
The female sex organs
(68, 69)
Reproduction (72, 73)
Pregnancy and birth
(76, 77)
Menstruation (70, 71)

The placenta

The fetus obtains oxygen and nourishment from the mother's blood via the *placenta*. There are hollow spaces in the placenta through which the mother's blood flows. In these spaces there are finger-like branches through which run the fetal blood vessels. These blood vessels have very thin walls, so that nutrients and oxygen can pass by diffusion from the mother's blood to the fetal blood and carbon dioxide and waste from the fetus to the mother, without the fetal blood coming in direct contact with that of the mother.

2nd-8th Week
The embryonic period

Part of the endometrium grows over the cluster of cells in order to protect them. The whole body continues to produce hormones. This prevents any new ovulation and so prevents the mucous membrane from being shed by menstrual bleeding.

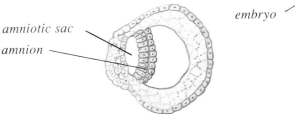

amniotic sac

amnion

Then two fluid-filled spaces are formed within the cluster of cells. The layer of cells between the two spaces is called the *embryo*. After 8 weeks it is called a fetus. One of the hollow spaces expands to become the amniotic sac which envelops the fetus, the other disappears. The fluid in the amniotic sac is called *amniotic fluid* and the membrane the *amnion (chorion)*.

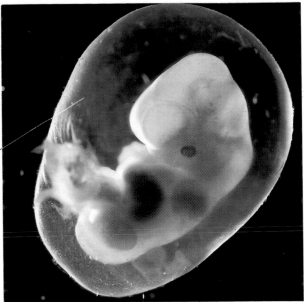

embryo

embryo at 6 weeks, 1.5 cm.

Most of the organic systems are formed during the embryonic period. At the end of the sixth week, the embryo is about 1.5 cm long.

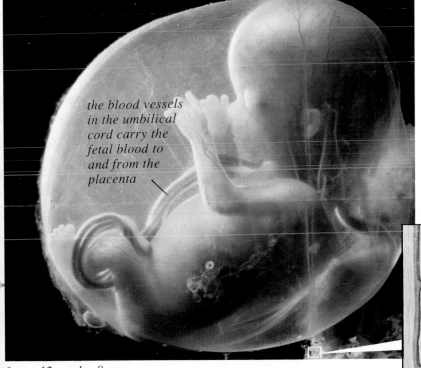

the blood vessels in the umbilical cord carry the fetal blood to and from the placenta

fetus, 12 weeks, 8 cm

The placenta produces progesterone, oestrogen and *HCG* (*Human Chorionic Gonadotrophin*) hormones. In the last six months of pregnancy, the placenta produces so many hormones that hormone production for the whole body is taken over by the ovaries.

9th-38th Week
The fetal period

In the fetal period the new individual grows and matures. When it is 25 weeks, it is ±35 cm long and weighs ±1 kg. At this stage the fetus is, with a bit of help, viable. It is, however, only fully developed after 38 weeks.

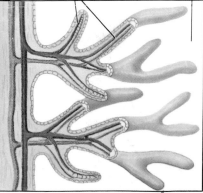

branches with fetal blood vessels

spaces filled with mother's blood

part of the placenta (greatly magnified)

Pregnancy and birth

Pregnancy lasts for about 40 weeks. During this time the fetus develops until it is capable of an independent life outside the uterus.

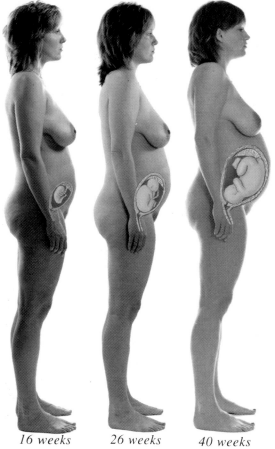

16 weeks   *26 weeks*   *40 weeks*

Pregnancy

The first sign that a woman is pregnant is the failure menstruation to occur. A few days later it is possibl detect the hormone HCG in the urine by means laboratory tests. This is a sure sign of pregnancy, si HCG is produced in the placenta.

Pregnancy lasts on average for 40 weeks and is cal lated from the first day of the last menstruation, wh means about two weeks before the actual fertilizat took place. The fetus then has 38 weeks in whic develop. In some cases the fetus is not viable. Suc pregnancy usually terminates with a *miscarri (spontaneous abortion)*.

To start with the bodyweight increases by 1 or 2 kg, not sufficiently to be able to see that the woma pregnant. Some women suffer from sickness, parti larly in the morning. As a rule the sickness discontin as the pregnancy progresses.

Then the bodyweight gradually increases, often by much as 12 kg. The uterus expands with the fetus increases in weight and volume, from 50 gr and 3 m 1 kg and 4 litres. The volume of blood and the circu tion increase because the placenta needs blood. H way through the pregnancy, movements of the fe can be felt. The breasts become enlarged and capable of producing milk. In the second half of pregnancy, a thick yellow fluid is often produ by the nipples. This fluid is called *colostrum*.

See also:
The female sex organs (68, 69)
For more about the fetus and the placenta, see: Embryo and fetus (74, 75)

About reflexes and epidural anaesthesia, see: The peripheral nervous system (36, 37)

Breast feeding

After the birth, the hormone *prolactine* from the pituitary gland initiates milk production in the mammary glands. The milk starts flowing after 2-5 days. This occurs through the response reflex. When the child begins to suck on the mother's nipple, nerve impulses from the nipple are sent to the nerve centre in the brain. This stimulates the secretion of the hormone *oxytocin* by the pituitary gland, which in turn ensures that the milk rushes into the breasts. It also stimulates the production of prolactine which maintains milk production.

The mother's milk is so constituted that all the child's nutritional requirements are met. The milk also contains *antibodies* which protect the child against infection. If the woman is not breast-feeding, milk production ceases after a few days.

The birth

A number of signals indicate the impending birth. *Birth pains* (*contractions*) very similar to menstrual pains occur with a certain regularity. The pains occur because the hormone oxytocin stimulates the smooth muscular tissue in the uterus wall to contract. When the contractions become regular and the intervals between them are less than 10 minutes, this indicates that the birth is approaching. When the fetal membranes burst, the amniotic fluid flows out.

Dilatation

The *dilatation* phase varies considerably (from 2 to 20 hours) but often takes longest in women having a first child. During this period the cervix expands and the pains increase in intensity and frequency.

Expulsion

Expulsion lasts from the moment complete dilatation is reached until the birth. This can last between half an hour and two hours. The woman can use her abdominal muscles (*abdominal pressure*) to press down and together with the contractions this is sufficient to expel the child from the uterus, down through the vagina to the outside world.

The newborn baby

With the baby's first cry, the lungs fill with air. The child therefore no longer has any need of oxygen from its mother. The blood vessels to the placenta via the *umbilical cord* are tied off, after which the umbilical cord can be cut.

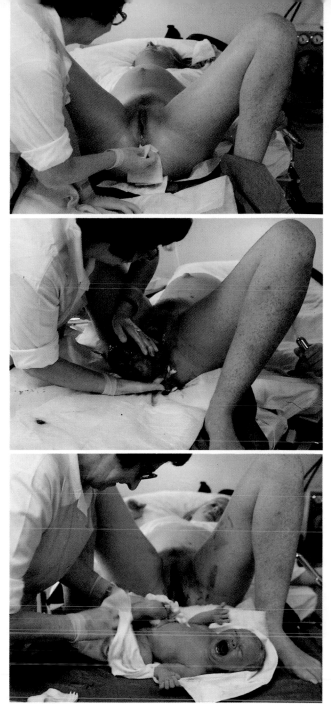

After-birth period

The time between the birth of the child and the expulsion of the placenta is called the *after-birth period*. As a rule the placenta is expelled within 15 minutes by the contractions of the uterine muscles.

Post-natal period

During the first few weeks after the birth, the uterus shrinks back to its normal size. There is often some discharge during this period. When the woman is not breast-feeding, menstruation starts again after about 6-8 weeks. If she is breast-feeding, then it recommences somewhat later.

Diseases   Smoking during pregnancy or lack of oxygen during childbirth can have harmful effects.

Heredity and environment

The characteristics which we develop are determined by genes which we get from our parents and by the environment in which we grow up.

Genetics

A new individual is created when a sperm with 23 chromosomes fuses with an ovum with 23 chromosomes. All cells in the new person have chromosomes which are copies of these 46 chromosomes (23 pairs).

Thus the chromosomes come in pairs. In 22 pairs the chromosomes are apparently identical. The 23rd pair consists of the sex chromosomes which do not have to be alike.

The chromosomes consists of pairs of linked genes. Identical chromosomes also have the same sort of genes. One pair of genes always influences the same characteristic, for example the colour of the eyes or the skin.

For more about skin colour, see:
The functions of the skin (56, 57)

About blood types, see:
The blood (22, 23)

See also:
Chromosomes and cell division (12, 13)
Reproduction (72, 73)

Genes and heredity

Although pairs of genes influence the same characteristics, this does not mean that the two genes are always completely alike. Sometimes the characteristic which finally dominates lies precisely in the middle between the characteristics which the genes would have created individually. Sometimes, one gene is 'stronger' than the other. We call the strong genes *dominant*, the weak ones *recessive*. One example is the ability to taste a particular substance. The gene which makes someone able to taste the substance dominates the others. When we get that dominant gene from one of our parents, we are able to taste the substance.

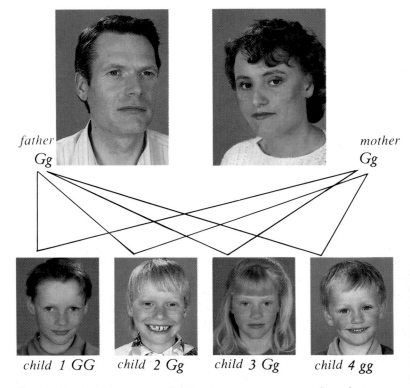

father Gg   *mother* Gg

child 1 GG   *child 2 Gg*   *child 3 Gg*   *child 4 gg*

G – is the dominant gene which ensures we can taste the substance.
g – is the recessive gene. With this we are unable to taste the substance.
Father, mother and children 1, 2 and 3 can taste the substance, but child 4 cannot.

Sex

In women the sex chromosomes are similar to each other. A female cell therefore has two X-chromosomes. In men, the sex chromosomes are not similar. A male cell contains one X- and one Y-chromosome. When in the woman the sex cells (ova) are created by cell division, the sets of chromosomes are halved. Each ovum contains therefore 23 chromosomes, of which one is always an X-chromosome. When the sperm cells are created, one half has an X-chromosome and the other half a Y-chromosome.

On fertilization, when the sperm has an X-chromosome, all the cells in the new individual have two X-chromosomes and become a girl. If the sperm which fertilizes the ovum has a *Y-chromosome*, the new individual will be a boy.

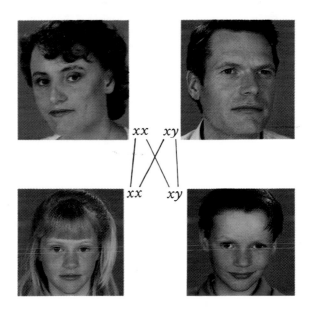

Environment and heredity

The majority of characteristics are not only determined by genes, but also by environmental factors. For example, we inherit from our parents many genes which influence our height. When people have tall parents, the probability increases that they themselves will be tall. But body growth also depends on a correct and sufficient supply of nutrients.

Many of the characteristics which are controlled by the brain are determined in the same way. Musicality is one example of this. With musical parents, the chances increase that the children will also be musical. Musicality is, however, also dependent on whether music is played and enjoyed in that family's environment.

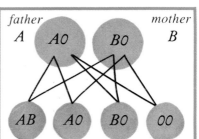

Blood groups

Genes of blood groups A and B do not dominate each other. An A gene and a B gene provide blood group AB. Both are, however, dominant in respect of blood group 0. Both the combination A0 and AA gives blood group A.

Diseases	Recessive diseases, haemophilia, muscular dystrophy, phenylketonuria (PKU).

Ageing phenomena

A series of interrelated changes occur in the human body during the lifespan. These gradual changes determine infancy, childhood, puberty, adulthood and old age. The physical capacity of the body is at its strongest between the ages of twenty and forty. After that the body slowly becomes weaker. With old people, the ageing phenomena are so clearly apparent that we talk of infirmities. Yet most healthy elderly people should regard old age as something positive, not least with regard to quality of life.

See also:
Sight (40, 41)
Hearing (38, 39)
The skin (54, 55)
The skeleton (58, 59)
The joints (60, 61)
The brain (34, 35)
The circulation (18, 19)
The heart (20, 21)
The sex organs (66-69)
Menstruation (70, 71)
The muscles (62, 63)
The defence mechanism (25)

Eyesight

As age advances, the lens becomes less supple, so that the eyes can no longer see sharply at short distances. Once past fifty, it is normal for people to start wearing spectacles for reading. The lens also becomes dimmer so that the elderly need more light in order to be able to read well.

Hearing

Many elderly people become hard of hearing. This is often due to calcification between the *ossicles* in the middle ear.

The skin

The skin becomes less elastic with age, so that wrinkles occur. A poor blood supply means that wounds can occur more easily and heal less quickly than in young people.

The skeleton

The bones become more fragile with age (*bone brittleness* or *osteoporosis*) and there is a greater risk of bones being broken. This particularly applies to women after the menopause, since the change in the amount of sex hormones affects the bones.

Exercise

Changes in the skeleton, the joints and the muscles can be limited by regular physical exercise. Inactivity can also lead to other complaints, such as constipation.

The joints

Wear and tear and calcification of the articulated joints, making them stiff and painful, often occur with the elderly. We call this *arthrosis*.

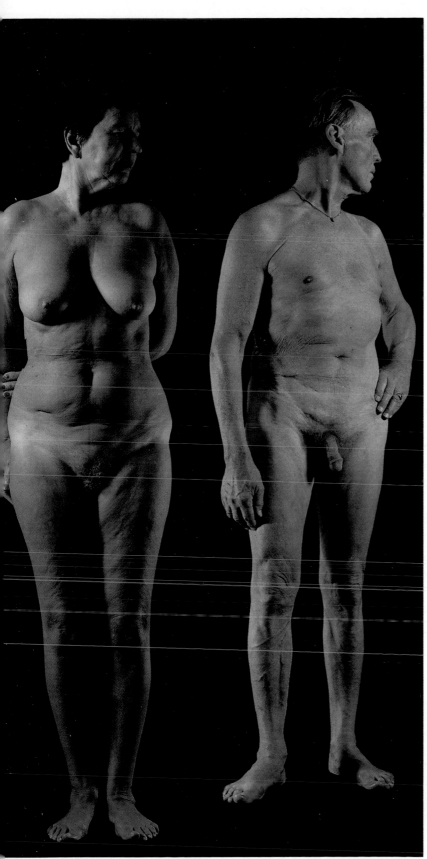

The brain

Ageing phenomena occur in the brain also. In most people, this is of little consequence, though they may be aware of increasing forgetfulness.

Some people may suffer from *senile dementia*. In some cases this leads to the elderly person no longer having any awareness of time, place or people.

The circulation

Blood circulation deteriorates with age. This is caused, among other things, by calcium deposits in the blood vessels (*hardening of the arteries*, arteriosclerosis), which makes them narrower. A poor supply of blood can result in the organs of the body functioning less well.

The sex organs

Women may suffer from a number of complaints during and after the menopause, because the concentration of sex hormones deteriorates. Men do not experience anything quite like the menopause, but the quantity of male sex hormones also decreases. Many men experience discomfort on urinating because the prostate has enlarged and is pressing on the urethra.

The muscles

Muscles decrease in size and lose their power and flexibility.

The defence mechanism

The body's defence against infectious diseases weakens over the years. Therefore many elderly people are more likely to suffer, for example, from pneumonia or other infections of the air passages.

Coordination of the body systems at rest

The body systems interact and are dependent on each other. Good interaction is a precondition for the normal functioning of the human body. Together, all our body systems form one large, well-regulated system, namely the human body. Even at rest, all the body systems participate in maintaining the body in working order.

For more about the interaction between the body systems, see also: Coordination of the body systems during physical exertion (84, 85)

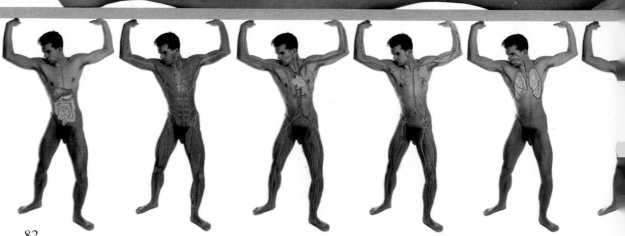

Even when we are resting, there is great activity within the body. Whilst we are eating and after we have eaten, the digestive system is working. Together with the muscular system it deals with the absorption, transportation and breaking down of nutrients, which are then carried round in the circulation system. The lymphatic system carries fats from the intestines to the main veins. The respiratory system ensures the intake of oxygen and the circulation system carries the oxygen from the lungs to all the cells in the body, so that the nutrients can be metabolized. The circulation system collects the waste materials from the cells and the respiration removes carbon dioxide from the blood. The kidneys remove excess water, salts and some waste materials. The portion of food which is not used, is exacuated by the digestive system via the faeces. Simultaneously with all these, the skin and the circulatory system control body temperature and the muscles and the skeleton keep the body upright. The sensory cells register everything that is happening both inside and outside the body and direct messages to the central nervous system. The central nervous system coordinates and controls the activity in all the body systems, in conjunction with the hormonal system. The last system, the reproductive system, is also active. In the reproductive organs, sex cells are constantly maturing.

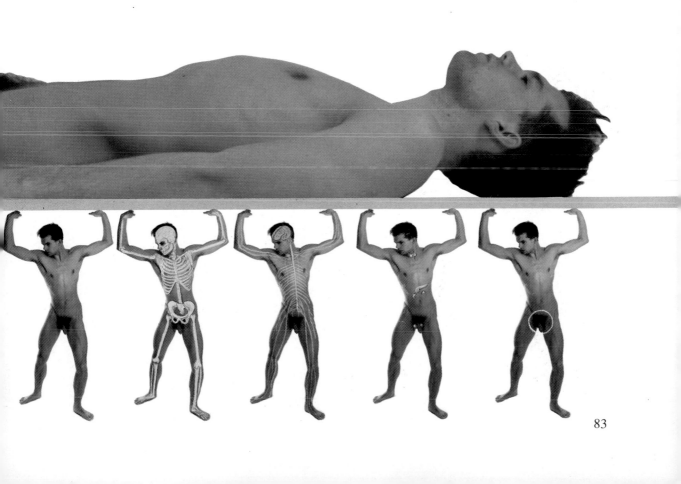

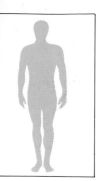

Coordination of the body systems during physical exertion

During physical work activity in the muscles increases. In order to allow the muscles to work as effectively as possible, the other body systems collaborate in order to increase the transport of oxygen and nutrients to the muscle cells.

For more about the work of the muscles, see:
The muscular system (62, 63)

See also:
Coordination of the body systems at rest (82, 83)

Under physical exertion, the skeletal muscles work so that we can move. The muscular system needs energy in order to be able to work and the energy requirement increases with exertion. During physical exertion, the muscle cells burn up a lot of energy and need a great deal of oxygen and nutrients.

The circulation system has to work harder to provide for these requirements and the heart beats faster and more powerfully than before. The increased blood circulation ensures that the waste material resulting from the metabolism in the muscle cells is quickly removed. The respiratory system accelerates and deepens breathing, so that we get more oxygen, whilst more carbon dioxide is exhaled at the same time. When the breathing and circulation systems are not capable of providing the muscle cells with sufficient oxygen, these cells are no longer able to completely metabolize the nutrients. Then *lactic acid* is formed as a waste product of the combustion.

In order to ensure as good a blood supply to the muscles as possible during heavy physical exertion, the circulation system adjusts the flow of blood. The heart and the skeletal muscles get extra blood, whilst, for example, the blood supply to the digestive system is reduced. Therefore, the activity in the digestive system is also reduced. This leads to the intestines not absorbing as much water and nutrients as usual, thus causing 'loose' stools. The nutrients, which are necessary for metabolism in the muscle cells, must consequently be drawn from the body's reserves. Changes in the hormonal system ensure that the stored materials are converted, so that they can be used by the muscles: for example glycogen is converted into glucose.

Heat is created by the considerable muscular activity during physical exertion. The skin prevents the body from becoming overheated by, among other things, exuding sweat which consumes body heat as it evaporates.

The nervous system directs the body and all the organic systems during the work so that everything is functioning correctly.

Activity increases in the various body systems during physical exertion.

Anatomical terms

It is not necessary to have a detailed knowledge of the names of the different anatomical terms in order to be able to understand the basics of the structure and functions of the body. However, it may be useful to know some of the anatomical terms.

The skeleton and the articulated joints

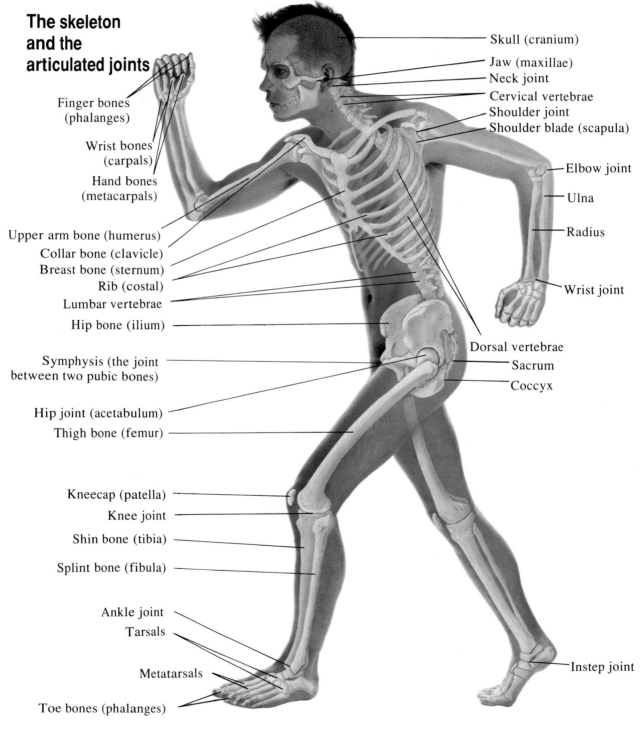

Skull (cranium)

Jaw (maxillae)

Neck joint

Cervical vertebrae

Shoulder joint

Shoulder blade (scapula)

Elbow joint

Ulna

Radius

Wrist joint

Dorsal vertebrae

Sacrum

Coccyx

Instep joint

Finger bones (phalanges)

Wrist bones (carpals)

Hand bones (metacarpals)

Upper arm bone (humerus)

Collar bone (clavicle)

Breast bone (sternum)

Rib (costal)

Lumbar vertebrae

Hip bone (ilium)

Symphysis (the joint between two pubic bones)

Hip joint (acetabulum)

Thigh bone (femur)

Kneecap (patella)

Knee joint

Shin bone (tibia)

Splint bone (fibula)

Ankle joint

Tarsals

Metatarsals

Toe bones (phalanges)

Muscles

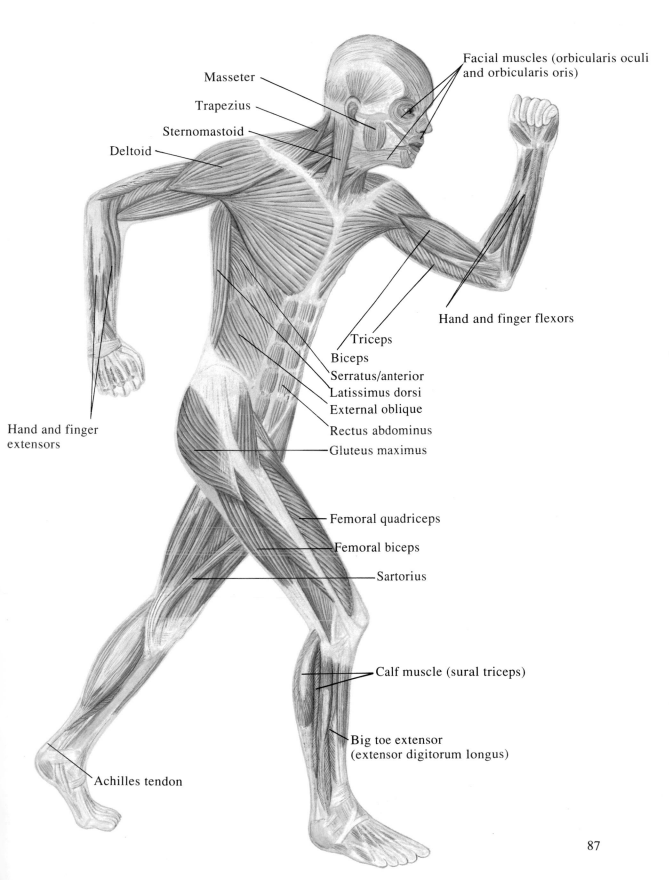

Masseter

Trapezius

Sternomastoid

Deltoid

Facial muscles (orbicularis oculi and orbicularis oris)

Hand and finger flexors

Hand and finger extensors

Triceps

Biceps

Serratus/anterior

Latissimus dorsi

External oblique

Rectus abdominus

Gluteus maximus

Femoral quadriceps

Femoral biceps

Sartorius

Calf muscle (sural triceps)

Big toe extensor (extensor digitorum longus)

Achilles tendon

Blood vessels
(the greater circulation)

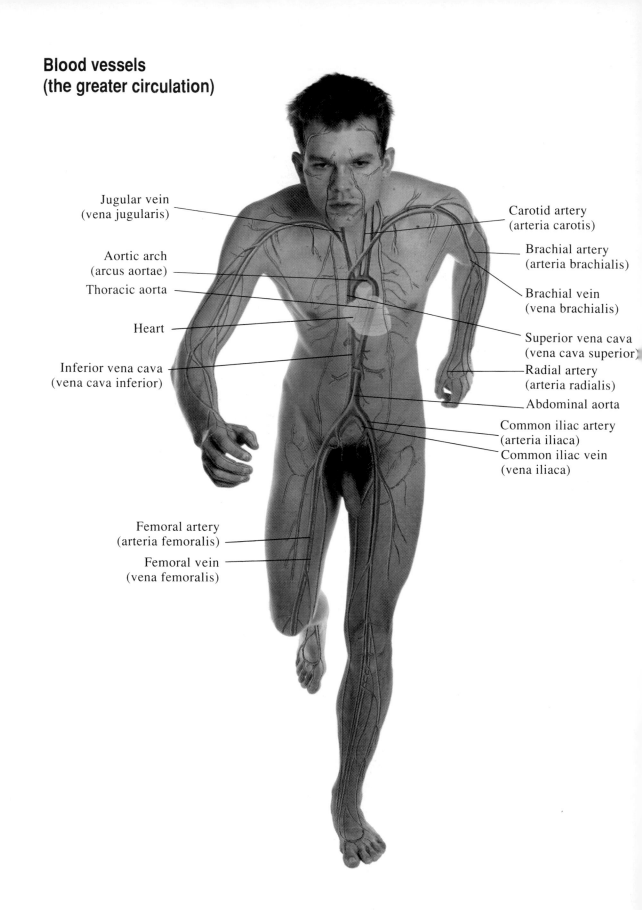

Jugular vein
(vena jugularis)

Aortic arch
(arcus aortae)

Thoracic aorta

Heart

Inferior vena cava
(vena cava inferior)

Carotid artery
(arteria carotis)

Brachial artery
(arteria brachialis)

Brachial vein
(vena brachialis)

Superior vena cava
(vena cava superior)

Radial artery
(arteria radialis)

Abdominal aorta

Common iliac artery
(arteria iliaca)

Common iliac vein
(vena iliaca)

Femoral artery
(arteria femoralis)

Femoral vein
(vena femoralis)

Nerves

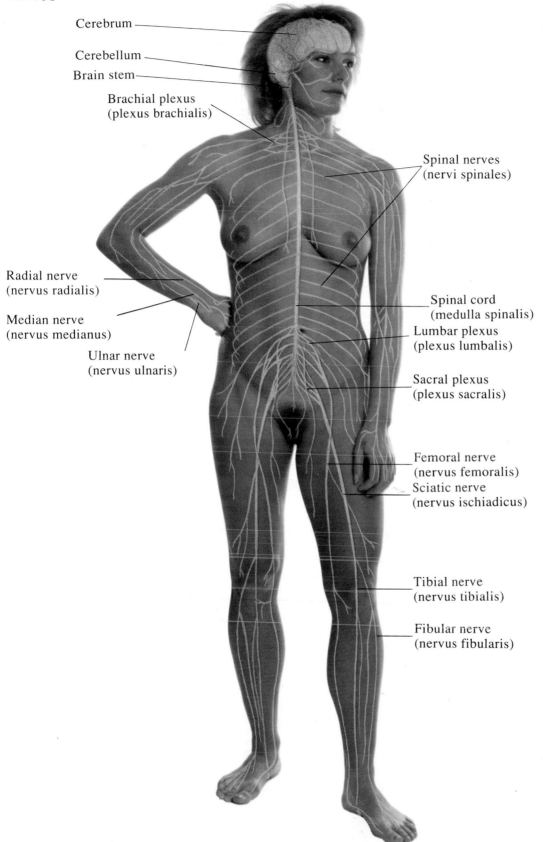

Cerebrum

Cerebellum

Brain stem

Brachial plexus
(plexus brachialis)

Spinal nerves
(nervi spinales)

Radial nerve
(nervus radialis)

Median nerve
(nervus medianus)

Ulnar nerve
(nervus ulnaris)

Spinal cord
(medulla spinalis)

Lumbar plexus
(plexus lumbalis)

Sacral plexus
(plexus sacralis)

Femoral nerve
(nervus femoralis)

Sciatic nerve
(nervus ischiadicus)

Tibial nerve
(nervus tibialis)

Fibular nerve
(nervus fibularis)

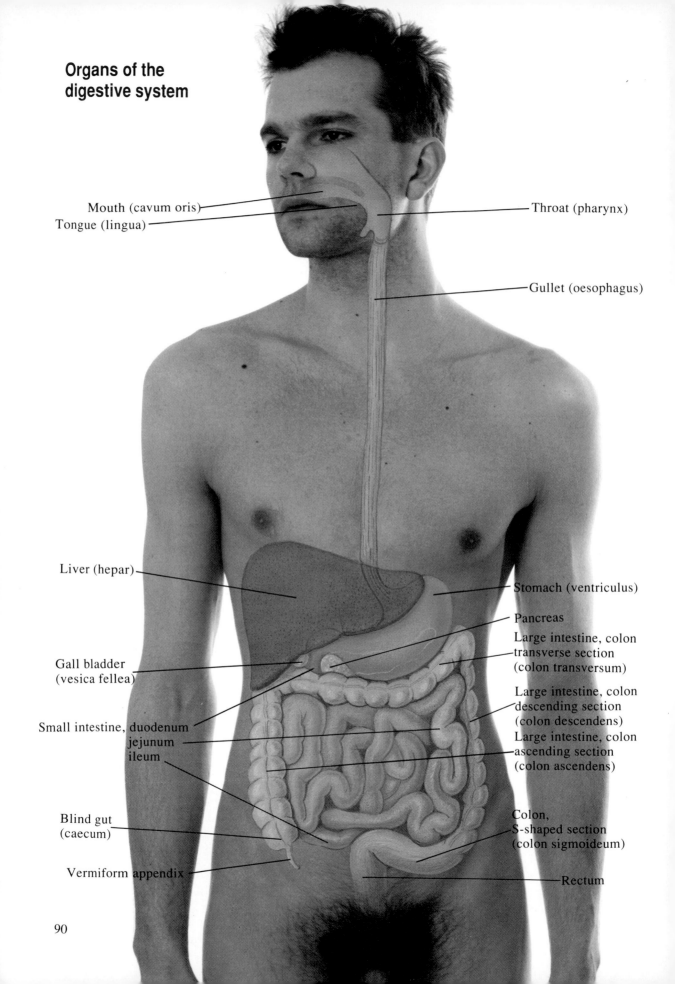

Organs of the digestive system

Mouth (cavum oris)

Tongue (lingua)

Throat (pharynx)

Gullet (oesophagus)

Liver (hepar)

Stomach (ventriculus)

Pancreas

Large intestine, colon transverse section (colon transversum)

Gall bladder (vesica fellea)

Large intestine, colon descending section (colon descendens)

Small intestine, duodenum jejunum ileum

Large intestine, colon ascending section (colon ascendens)

Blind gut (caecum)

Colon, S-shaped section (colon sigmoideum)

Vermiform appendix

Rectum

90

Tables

Processing of food in the digestive system

Organ	Digestive process
Mouth	Divided up and chewed (masticated) Mixed with saliva
Throat and Gullet	Swallowed
Stomach	Kneaded and diluted with gastric juice to a pulp Proteins broken down Bacteria killed
Small Intestine	Pancreatic juice, bile and digestive enzymes from the tiny glands in the intestinal wall are added Proteins, carbohydrates, fat and nucleic acids are broken down Nutrients, salts, minerals, vitamins and water are absorbed
Large Intestine	Water, salts and minerals are absorbed or evacuated The remainder of the waste materials and dead bacteria form the faeces

Hormones

Hormone	Secreted by	Influences	Effect
Thyroid stimulating hormone (*TSH*)	Pituitary, front lobe	The thyroid	Increases secretion by the thyroid hormone
Adrenal cortex stimulating hormone (*ACTH*)	Pituitary, front lobe	The adrenal cortex	Increases secretion by the adrenal cortex hormone
Growth hormone (*GH*)	Pituitary, front lobe	The whole body	Promotes body growth
Follicle stimulating hormone (*FSH*)	Pituitary, front lobe	The sex glands	Growth of rudimentary ova Sperm production
Luteinizing hormone (*LH*)	Pituitary, front lobe	The sex glands	Stimulates ovulation Stimulates production of male sex hormones
Prolactine (*PRL*)	Pituitary, front lobe	Mammary glands	Growth and milk production
Antidiuretic hormone (*ADH*)	Pituitary, rear lobe	The kidneys	Reduces amount of urine discharged
Oxytocin	Pituitary, rear lobe	Smooth muscular tissue in the uterus and breasts	Birth pains, ejection of milk, uterine contraction
Thyroid hormone (thyroxine)	Thyroid gland	The whole body	Raises metabolism
Calcitonin	Thyroid gland	Calcium balance	Reduces calcium content of the blood
Parathyroid hormone	Parathyroid glands	Calcium balance	Increases calcium content of the blood

Hormone	Secreted by	Influences	Effect
Aldosterone	Adrenal cortex	The kidneys	Stimulates the absorption of sodium Stimulates the excretion of potassium
Cortisone	Adrenal cortex	The whole body	Influences metabolism Concerned with crisis reactions. Slows down inflammatory reactions
Adrenaline	Adrenal medulla	The whole body	Prepares the body to meet crises, increases blood sugar, raises blood pressure
Noradrenaline	Adrenal medulla	The whole body	Prepares the body to meet crises, raises blood pressure
Insulin	Pancreas	The whole body	Lowers blood sugar level (glucose)
Glucagon	Pancreas	The whole body	Raises blood sugar level (glucose)
Oestrogen	Ovaries (and adrenal cortex)	The whole body	Controls sexual maturity, reproduction, and the development of sexual characteristics
Progesterone	Ovaries Placenta	The sex organs	Stimulates secretion in the endometrium, protects pregnancy
Androgens, including testosterones	Testicles (and adrenal cortex)	The whole body	Controls sexual maturity, reproduction, and the development of sexual characteristics

Index